Ear Function in SSRI Withdrawal:

A Comparison with Symptoms of Known Ear-Related Syndromes Treatable with High-Frequency Music

Laurna Tallman

Northern Light Books
2012
Revised
2018

Library and Archives Canada Cataloguing in Publication

Tallman, Laurna
Ear function in SSRI withdrawal : a comparison with symptoms of other ear-related symptoms / Laurna Tallman.

Includes bibliographical references.
ISBN 978-0-9811672-2-0

1. Sound--Therapeutic use. 2. Music therapy. 3. Antidepressants--Side effects. 4. Affective disorders. 5. Developmental disabilities. I. Title.

RZ999.T35 2012 615.8'5154 C2012-905948-X

Abstract:

The author describes a spectrum of disorders—dyslexic syndrome, chronic fatigue syndrome, and bipolarity and schizophrenia induced by illicit drug use—that were shown to have their etiologies in the right middle ear when they diminished and/or disappeared following stimulation of the right ear with high-frequency sound.

The author considers data volunteered by 107 persons (1) discontinuing SSRIs and other psychoactive medications and/or (2) with residual symptoms after cessation. Of those subjects, 98% and 91%, respectively, have symptoms consonant with severe right middle and inner ear damage. All suicidal subjects in both categories have symptoms of severe middle ear and inner ear damage. More than half of those with suicidal tendencies report symptoms of dysfunction in the vestibule, cochlea, or both parts of the inner ear. Among suicidal subjects reporting only one locus of inner ear symptom (cochlea or vestibule), twice as many subjects reported vestibular symptoms. Whether in withdrawal or when dealing with residual symptoms, more than 50% of the people at either stage of discontinuation who show moderate symptoms of middle ear damage contemplate or attempt suicide. Sixty-two per cent of the subjects in SSRI withdrawal and 63% of persons with residual symptoms report having problems with the inner ear, either the vestibule (46% and 42%) or cochlea (39% and 35%) or both (21% and 22%).

In view of the spontaneous recoveries and partial recoveries of these subjects and the similarity of their symptoms to those improved or healed by focused listening to high-frequency music, it seems likely that listening to high-frequency music with headphones could greatly accelerate the SSRI withdrawal and recovery process. Because many of the presenting problems for the prescribing of SSRIs are syndromes that already have been shown to be responsive to high-frequency music, including suicidal depression and autism, interventions with high-frequency music, such as focused listening, singing, playing a high-frequency stringed instrument (violin or guitar), the Tomatis Method, or Bérard AIT may heal those syndromes simultaneously with the symptoms caused by SSRIs. Bérard's finding that specific audio profiles are associated with suicidal depression, milder depression, and aggression, affirm the author's hypothesis that not only mental illnesses, but such conditions as discontinuation syndrome describe a spectrum that might be calibrated to audio-processing deficits in hertz in either the right, the left, or both ears and that might be treated with focused listening to high-frequency music rich in the frequencies the patient cannot perceive normally. Appendices include summaries of the symptoms under discussion: dyslexic syndrome, chronic fatigue syndrome, bipolarity and schizophrenia, and SSRI withdrawal syndrome.

Keywords: SSRI discontinuation; SSRI withdrawal; antidepressants; psychoactive drugs; chronic fatigue syndrome; dyslexic syndrome; bipolarity; suicidal depression; schizophrenia; music therapy; Focused Listening, Tomatis Method; Bérard's AIT.

Northern Light Books
240 Beaver Creek Road
Marmora, Ontario, Canada K0K 2M0
www.northernlightbooks.ca

Ear Function in SSRI Withdrawal:
A Comparison with Symptoms of Known Ear-Related Syndromes Treatable with High-Frequency Music

Introduction: SSRIs and Other Psychoactive Medications
The positive effect of music on mood has been recognized for millennia but, until now, has not been explained. Like the neuroscientists and psychiatrists who have been running around in chemical circles for a century, most people have been lured into thinking about the chemistry of the brain without understanding the physics of sound in neurology. Even those opposed to the use of SSRIs ruminate about serotonin and dopamine in the brain.[1] The ignorance of doctors and psychiatrists is widespread as regards the ear's control of the body's neurology and of its organs and systems. Such knowledge has been available to researchers for 70 years, since Dr. Alfred Tomatis learned that the ear controls the voice (larynx muscles). Perhaps it was more difficult in the 1950s for research discoveries to cross the Atlantic Ocean. Soon afterwards, the rise of pharmaceutical "easy answers" in psychoactive medications, such as lithium, benzodiazepines, and SSRIs including Paxil, Zoloft, Prozac, Lexapro, Celexa, and Effexor, obviated the search for the origins of aberrant behaviour. Neuroscience became completely preoccupied with the chemistry of the brain while neglecting sensory input to the brain. Tomatis and his colleague Dr. Guy Bérard founded organizations for the treatment of people with the syndromes they had discovered were responsive to high-frequency music stimulation with their patented electronic equipment for filtering out the lower frequencies from the music. Their audio training enterprises were only rarely drawn into mainstream medical practice.

We now have a rationalization of *harmful psychiatric patient care* that is like walking into a dark house knowing about the chemistry of fire, but not about the physics of electricity. Neuroscience and psychiatry are trying to illuminate the house by setting fire to the flammable materials. They may be able to strike a spark, but the house will suffer and the homeowner may suffocate.[2] Alternative strategies may (or may not) open curtains to let in more light (e.g., vitamins and exercise) or call in extraordinary light sources (e.g., hyperbaric oxygen). Few realize they could use high-frequency music to turn on the lights. Music is not a soothing way of surviving in the dark. It is *the* form of physical energy that turns on the main neurological switch. We are familiar with light-sensitive electrical gadgets. The ear is a sound-sensitive gadget, composed of muscle, bone, and other tissues, that starts to convert sound energy into electrical energy in the middle ear's connection to the nervous system and that processes sound into the inner ear, which further transforms the sound energy it delivers to the brain. If you tune in to music by singing and by listening to high-frequency music, you will activate the main switch, the existing wiring will work, the lights and appliances will turn on, and you won't have to struggle to survive in smoke-filled rooms damaged in other ways by the chemical fires. Music is the obvious, natural way to repair and maintain a nervous system that carries electrical impulses at about the speed of sound.[3]

High-frequency music exercises the tiny stapedius muscle in the middle ear, making it more fit to control how sound reaches the cochlea and vestibular labyrinth to profoundly impact the body's neurology, especially the brain stem and the temporal lobe in the left cerebral hemisphere. Connected by a fibre of the vagus network to many body systems, it influences the body from the hormonal secretions in the brain to the end of the digestive system. If the muscle is too severely damaged, it may not be recoverable. However, our son Daniel recovered from at least eight drug-induced episodes of schizophrenia, which at first were heavily medicated. His recovery provides reason for hope for those people similarly damaged by pharmaceuticals. I have seen persons in a drunken stupor respond within a couple of minutes to the stimulus of high-frequency music by showing greater alertness and rationality. I have known bipolar persons to become calm, rational, and more concerned about others and to become able to sustain non-bipolar behaviour through exposure to high-frequency sound. The spontaneous recoveries and partial recoveries of the subjects of this study, presumably without music stimulation—although no one asked that question—are further reasons for optimism. It seems likely that high-frequency music could greatly accelerate the SSRI withdrawal and recovery process. When you consider that many of the reasons for the prescribing of SSRIs are symptoms listed in Appendix A that already have been shown to be responsive to music, music may heal those symptoms when they are caused by SSRIs. Because Daniel's schizophrenia was drug- and medication-induced, my discoveries bridge gaps in the work of Tomatis and of Bérard.[4] Daniel's many withdrawals from street drugs and prescription drugs, mostly under my supervision, also have given me insights into SSRI discontinuation (withdrawal) syndrome.

Assaults on the Ears

Daniel, born in 1980, had shown symptoms of dyslexic syndrome from infancy.[5] My professional expertise as an educational researcher and as a senior editor of texts for the teaching of reading, which I brought to the analysis and support of his learning problems, was confirmed many times by tests done through his schools. State of the art educational theory held that dyslexia, deemed "minimal brain damage," was incurable, but that learning could be "supported." Terminology now in use has the same prognosis, so ADD kids often are drugged.

Following a flu epidemic in 1988, three other members of our family developed such a complex array of symptoms that no one of us could imagine the others suffered from the same illness. After extensive testing to rule out other possible causes of our symptoms each of us was diagnosed as having chronic fatigue syndrome (CFS).[6] Chronic fatigue syndrome is a diagnosis of exclusion. Shortly after our diagnoses in 1994 by our general practitioner, two of us were seen in Ottawa by the associate of Canada's leading expert on CFS, Dr. Byron Hyde. He was one of a few doctors like Dr. Jay Goldstein of Anaheim, California, who were conducting research on the syndrome. Our family doctor's diagnosis was confirmed in Hyde's clinic. Shortly thereafter, I obtained Goldstein's "Checklist,"[7] which was helpful in showing how the same illness affected individuals differently. Of 56 symptoms listed by Goldstein only five of the more common symptoms and six of the infrequent or rare symptoms had not been experienced by one or more of us; and we had experienced several other symptoms fairly typical of the condition according to the newsletter of Hyde's Nightingale Foundation. Like Daniel's dyslexic syndrome, CFS was deemed incurable. Various supportive measures were offered to us by physicians including SSRIs (selective serotonin reuptake inhibitors), which we used briefly or not at all.[8]

In 1997, Daniel experienced a sustained psychotic break and was heavily medicated in a general hospital and in an Ontario psychiatric hospital. He was diagnosed as schizophrenic with the prognosis from one of eight psychiatrists who saw him, "once a schizophrenic, always a schizophrenic." About that time Daniel disclosed that he was using alcohol and drugs extensively and addictively.

Healing Dyslexic Syndrome, CFS, and Schizophrenia with Music

Daniel was 16 when he was healed of his life-long dyslexic syndrome during 10 days of two-hour listening sessions at the Tomatis Method[9] Listening Centre in Toronto, Canada, in 1997.[10] Our family members with CFS symptoms (which ranged from moderate to severe) were healed or found our symptoms greatly improved through binaural ear stimulation with electronically filtered music listened to through headphones provided by The Listening Centre in Toronto or in their portable LiFT® device, (in 1997, 1998, and 2002).

Goldstein identifies some CFS symptoms as "attention deficit disorder," although he does not mention the full range of symptoms that characterize dyslexic syndrome (see Appendix A). Therefore, at least some of Daniel's symptoms of dyslexic syndrome overlapped with some aspects of our CFS syndrome, although we had not thought of our CFS or his ADD from that perspective. Tomatis's diagnosis and healing of CFS—although he does not use that term but describes some of the symptoms of debilitating mental and physical fatigue and irritability[11]—was famously brought to 70 French monks when he prescribed that they should return to singing. Their daily chanting had been dropped with other changes in their religious order post-Vatican II. They also used his Electronic Ear® for listening to music filtered to deliver only the higher frequencies.[12] I had no expectation that listening to electronically filtered music could change anything about me, yet, after about 7½ hours of listening to filtered high-frequency music (Mozart violin concertos), spread over 4 days, I was restored in an instant to normal alertness with normal energy and a cessation of virtually all my CFS symptoms. It felt to me as though a switch had been thrown on in the centre of my head. My normal energy returned in a rush. My muscles probably did not return fully. I was post-menopausal and had experienced significant muscle wasting and some weight gain. But I was able to regain fully normal, active life *immediately*. Two years later, I relapsed and eventually learned that my healing had to be maintained with regular singing and listening to high-frequency music whenever my ear came under some kind of assault, such as by a cold virus or some other kind of throat infection.

Within days of Daniel's dramatic healing from dyslexia he returned home and experienced a prolonged psychotic break, following which he was declared hopelessly schizophrenic, which also is a diagnosis of exclusion. One of his psychiatrists told me the Tomatis Method listening intervention had *caused* his schizophrenia but offered no explanation as to *how* such a thing could have happened. Could it have, when it had cured his dyslexia and when I had experienced a tremendous healing of CFS by the same method? The direct results of his music intervention had been an astonishing normalizing of his behaviour. How could that have turned into a "mental illness"? Our family was thrown into a vortex of confusion, anxiety, and unusual stress.

For the next 10 years of his schizophrenic illness we tried to help Daniel come to terms with his addictions. I referred to his acute episodes of insanity as "psychosis," as I remained doubtful of a diagnosis that the psychiatrists themselves could not explain or define. I gradually learned strategies that improved Daniel's mental condition, especially during the first three years, by working with his doctor to cease or reduce his poly drugging to one medication at a minuscule fraction of the dosage he was given in hospital. I noticed many things about his symptoms during his episodes of schizophrenia and I tested his peculiar, fluctuating levels of cognition, finding them consistent at two-minute intervals to make a four-minute cycle. I tested him in separate schizophrenic episodes years apart, obtaining the same results. Daniel said several times at the onset of acute psychosis, "I'm dyslexic again." Still, I did not make the connection between his dyslexic *hearing* and his episodes of severe "mental" illness. I thought he meant he was experiencing two different illnesses concurrently. Because Daniel tapered off his antipsychotic medication entirely when he recovered from an episode of schizophrenia (he improved greatly several times), his behaviour between acute episodes was fairly normal and showed gains in maturation, except for the symptoms produced by his use of cannabis and hallucinogens that would gradually build towards another episode of acute schizophrenia.

Only after Daniel became schizophrenic twice again (2005–6; 2008) and was healed of those symptoms at home by Focused Listening—listening to music with only the right earpiece of his headphones to ordinary CDs of classical violin music—did I see that some of the symptoms on Goldstein's "Checklist" include dyslexic syndrome symptoms milder than our CFS symptoms and other symptoms in a range more severe than our experience of CFS, namely, depression, obsessive-compulsive disorder (OCD), bipolarity, and schizophrenia, that characterized Daniel's drug-induced, severe, "mental" illness.

In 2008, I discovered that Daniel's dyslexia and his schizophrenia *both* were characterized by shifts in cerebral hemispheric dominance—that in schizophrenia was so extreme as to create non-dominance and virtually no cerebral integration. My study of the anatomy of the ear showed me that dysfunction of the ear was responsible not only for those two distinctive patterns of behaviour but for other mental conditions. Dyslexia and schizophrenia were not concurrent, because they are mild and severe versions of the same neurological phenomenon. It was evident from Daniel's shifts in behaviour patterns that any number of patterns of behaviour, loosely categorized by psychiatry as "mental," actually were caused by a dysfunctional ear. Through my study of neurology, I became able to explain how all of those aberrant, unhealthy, abnormal, or "mental" behaviours in Daniel were *changed* through exposure to high-frequency music focused on his right ear. With the exception of Daniel's most extreme schizoaffective and schizophrenic behaviours, those are the same kinds of changes brought about daily in hundreds of Tomatis Method and Bérard AIT listening programs using binaural methods. My distinctive therapy of Focused Listening revealed a neurological paradigm that doctors were not aware of: The right ear makes the left-brain dominant over the right-brain in their integrated activities. The flow of sound energy has a necessary *direction* through the brain and nervous system.

Music heals the body and normalizes behaviour because the cause of aberrant behaviour and much ill health is a right (or left, or both) middle ear muscle that cannot adequately process high-frequency sound. ***The corollary of that rule is that the normally healthy ear, equally, is the source of normal behaviour.*** The stapedius muscle controls the action of the stapes at the oval window, where high-frequency sound is transmitted to the inner ear. Tomatis, Bérard,[13] and I have shown that a wide range of aberrant behaviours can be altered if that muscle is stimulated by the music of violins. Tomatis's and Bérard's methods use electronically filtered music. I use ordinary CDs of violin music. In both techniques, music is focused onto the eardrum by headphones. Virtually all of the symptoms listed in Appendix A (that coincidentally includes most of the symptoms on Goldstein's "Checklist") have been shown by Tomatis, by Bérard, or by me to be ear-related and treatable with high-frequency sound.

SSRI Discontinuation and Residual Symptoms

Coming upon the Internet forum Paxilprogress, which listed, for the use of researchers, detailed symptoms of people in severe discontinuation (withdrawal) syndrome as a result of diminishing or ceasing their SSRIs and other psychoactive medications, I noticed that most of their symptoms were similar to the symptoms our family (and thousands of other people) had experienced from moderate to severe viral damage to our ears (CFS) and that Daniel had experienced from the traumatic effects of drugs, loud sound, and head concussion that had damaged his ears[14] (depression, obsessive-compulsive disorder, manic-depression, and schizophrenia). Furthermore, I had seen many of those symptoms repeatedly as withdrawal symptoms in Daniel (and in others) when he was diminishing or ceasing his intake of street drugs or of psychoactive prescriptions, or both. Psychoactive medications can cause the very symptoms they are intended to control[15]—a result I have observed many times and that is mentioned frequently by the subjects of this study,[16] and by the pharmaceutical companies, although less transparently.

I wanted to see if those discontinuation and residual symptoms could be sorted according to the specific parts of the ear that control them. I was guided by the anatomy taught by Tomatis in his writing and by my discoveries related to the dominance of the left cerebral hemisphere (left- brain) where rationality, coherent language, nuanced speech, and self-control operate. Some symptoms and behaviours pertain more to the cochlea (e.g., cognition and speech); some are more under the influence of the vestibule (e.g., balance and movement); and some are specific to the vagus network (digestion, respiration, cardiovascular function, the immune system, and certain hormonal secretions) that is connected by one of its fibres to the stapedius muscle. The stapedius muscle is attached to the stapes (stirrup) and controls its action at the oval window. The intricate neurology of those cybernetic neurological loops to brain and body comes back to the ear. It is the body's timing mechanism. The parts of the middle and inner ear act synchronously, under the stream of high-frequency sound. When the stapedius muscle(s) is impaired, all Hell breaks loose.[17]

Many people in withdrawal from SSRIs or experiencing residual symptoms display suicidal tendencies. Bérard discovered that suicidal depression is caused by hypersensitivity (hyperacusis) in the left ear (or, more rarely, the right ear) at 2 and 8 kiloHertz— the "2–8" profile.[18] His proof that high-frequency sound can normalize perception of sound at those frequencies does not provide an explanation for how strengthening one ear or the other ear rearranges the person's self-destructive thought process.[19] Certainly, Bérard's binaural treatment normalizes the stream of sound from the left ear to the right-brain. It enhances the stream of sound from the right ear to the left-brain, too. While I think stronger left-brain dominance is a factor in reduced suicidal tendencies, not enough information is provided in Bérard's brief vignettes to be certain. Thus, I have placed subjects with suicidal tendencies in a separate category. Bérard's proof that the etiology of severe depression lies in the dysfunction of the ear also proves the *physiological* nature of a major mental illness and refutes Tomatis's idea that people who don't want to hear stop listening intentionally by blocking the flow of sound through the ear. His findings refute notions in psychiatry and religion that blame suicidal people for "psychological" or "spiritual" failures. Such suicidal people are *physically impaired.*

My discovery of the effect on behaviour of losses of left-brain dominance permits me to assign symptoms causally to the middle ear that neither Tomatis nor Bérard appear to have recognized. Tomatis was not mistaken in thinking that other behavioural syndromes—serious mental illness—also had their etiology in the middle ear,[20] but he should not have listened to his colleagues in psychiatry. My research shows that the degree of volitional control by the left-brain of the stapedius muscle is a function of the tonus (strength and flexibility) of that muscle.[21] In fact, deliberate action is physically impossible when the brain's hemispheres are integrating slowly, just as mental or physical action is impossible when they integrate very slowly during normal sleep. Like the audio-phono loop in which the ear controls the voice, the relationship of the right ear with the left-brain is a cybernetic loop: the ear controls the brain by maintaining dominance in the left cerebral hemisphere through the ear's ability to adjust the flow of high-frequency sound to the left temporal lobe. Only when the left hemisphere is kept dominant with that energy stream is it possible for the person to learn to exercise control over behaviour prompted by the primal urges and by the emotions experienced in the right-brain. Among those forms of self-control that children learn gradually is the semi-conscious and unconscious control of the right stapedius muscle. It operates somewhat like the eyelid, which reacts spontaneously to shut out light but may also be lowered deliberately to shut out light. As the movement of the stapedius muscle cannot be seen, the child develops an inner sense of its spontaneous activity and learns to control it through feedback from the body's muscles that the middle ear muscles control. The stapedius is an extensor and controls the extensor muscles. The muscle of the eardrum is a tensor and controls the tensor muscles (Tomatis). The stapedius is also subject to socialization but less easily and more privately than control of the eyelid. You can teach a child to close her eyes. It is much more difficult to teach a child to close his ears and go to sleep, for example, or to open her ears and pay attention. That teaching has been made more difficult because we have not understood exactly how the ears are involved in those *self-regulating* activities.

Cerebral Integration

I had learned through personal experience and through research that similar behaviours or symptoms are produced by various assaults[22] on the ear:

1. chemicals (e.g., psychoactive drugs, drugs such as prednisone that weaken muscle, gases, anesthetics, certain antibiotics, caffeine, beverage alcohol)
2. loud sound; low-frequency sound
3. infections (viral, bacterial, yeast, mold)
4. trauma (falls, sporting and auto accidents, beating, extremes of atmospheric pressure, tumours)
5. metabolic and hormonal disturbances
6. oxygen deprivation (carbon monoxide poisoning, a difficult birth, inability to exercise, inadequate nutrition)

Members of our family learned eventually that if our ear muscles were not regularly exercised with sound (singing and listening to high-frequency music), or if the ear was subjected to some kind of assault, such as those listed above, the symptoms of CFS or of dyslexia or of schizophrenia could return. A range of schizoaffective-disorder symptoms (distinguishable from full-blown schizophrenia) were often evident in Daniel.[23] Those distressing behavioural syndromes, which lie on a spectrum between schizophrenia and dyslexia, were healed with high-frequency sound as he progressed towards normal brain function using Focused Listening. In other words, the symptoms of mental illness form a continuum or spectrum dependant on the level of fitness of the right ear's stapedius muscle. Another such spectrum was discovered by Bérard in the left-ear that shows specific audio-deficits related to a range of levels of depression. I have witnessed shifts back and forth along the right-ear spectrum in dozens of people. I discovered that all of those symptoms are related in one way or another to dominance diminished or lost in the left hemisphere of the brain. The right-brain also needs an unimpeded supply of sound energy for its needs, which include maintaining feelings of optimism.

The Ear and Behaviour

The role of the ear in hearing (the cochlea) and in balance (the vestibule) is widely known. Tomatis demonstrated additionally that the perception of frequencies by the right ear—measurable in Hertz and kiloHertz—determines the pitch of the voice: the Tomatis Effect (demonstrated to the French Academy of Medicine, 1957). He and Bérard proved further through their protocols for treating tens of thousands of patients and clients that stimulation of the middle ear's muscles (especially the stapedius muscle, according to Tomatis) with filtered music (high-frequency sound) can correct an array of symptoms associated with abnormalities in singing and speech—stuttering, lisping, lack of vocal intonation, the inability to hold a tune—but also the syndromes of dyslexia, moderate to severe chronic fatigue, moderate depression, suicidal depression, and the very severe symptoms of autism (infantile schizophrenia), although he was prevented from completing his work with the autistic children who were recovering with the Tomatis Method.

During Daniel's episodes of acute schizophrenia, I gradually gathered information about his behaviour that allowed me to discover that *alternating* cerebral influence was the defining symptom of his schizophrenia—as I could see, then, that it had been of his dyslexia to a lesser extent. Daniel's losses of left-hemisphere dominance explained his schizophrenic losses of rationality, of normal beliefs, and of self-control; his inability to act deliberately (as opposed to spontaneously); his inability to order words rationally or grammatically; and his loss of control over his emotions and the emotional prosody of his voice, among other symptoms[24] (see Appendix A). Similar losses of self-control are evident among the symptoms reported by those suffering SSRI discontinuation syndrome (Appendix B).

In the process of recovering from schizophrenia Daniel passed through distinctive behavioural syndromes: about three decreasing levels of severity of schizophrenia; then, bipolar I; bipolar II; obsessive-compulsive disorder (OCD); mild depression; dyslexia; and, finally, normal integration speeds under left-brain dominance. Once I had seen that progression, I looked at my records and saw that he had passed through those stages *in reverse order* and very quickly during the onset of full-blown schizophrenia. During his relapse in 2016, I was much more aware of this progression of phases that marked the collapse of his stapedius muscle(s) during his deterioration, which took weeks but escalated during the final days when manic pacing and uncontrolled verbalizing began. Those stages constitute many of the so-called "mental" illnesses that Tomatis was convinced were related to ear function.[25] As Bérard had measured the deficits of his depressed patients in "1–8" and "1.5–8" profiles and his suicidally depressed patients in a "2–8" profile on their audiograms, so each of Daniel's behavioural sets or syndromes must correspond to a characteristic audiogram curve that would have been measurable in Hertz if I had possessed the equipment to measure it.

That is not to say that volitional or semi-volitional control of the stapedius muscle was never within Daniel's capability. However, until his stapedius muscle had been exercised enough to make it sufficiently fit (tonic) to cause neurons to fire—thus joining its activity to the left-brain in a reciprocal pathway—he had neither the means of exercising control over that tiny muscle nor, therefore, over much of his other behaviour. Thus, the lack of conditioning of the stapedius muscle accounts for crying jags, fits of rage, outbursts of frustration, pressured and incoherent speech, and most other forms of uncontrolled behaviour. Not only in Daniel, but in any of us, losses of self-control are to some degree "psychotic." During childhood we learn gradually to control such behaviour according to social and cultural norms. Those gradations of ear function correspond to particular levels of self-control that can be learned via changes in the stapedius muscle felt through changes in the overall tension in the muscles of the body. The right ear to left-brain to larynx loop develops a cybernetic of cerebral dominance controlled by the ear that controls other behaviour. Or, as Tomatis wrote, "As we move towards effective listening uncontrolled behaviour occurs less often."[26]

Emotional Prosody

Self-control includes the capacity of the left, rational brain to attend (listen) to and respond with rational language using the right ear, while imbuing the voice with emotional meaning by accessing input from the left ear via the emotional right-brain. "Emotional prosody"—the infusions of appropriate emotion into tonally nuanced speech—is a contribution of the *left ear* to which the left temporal lobe attends subordinately and selectively while Wernicke's area in the left-brain responds by ordering words logically and grammatically to input from the right ear. These selective processes in the left-brain are possible because any given sound arrives from the left ear a fraction of a second later than the same sound arriving from the right ear. The neural pathway from right ear to left-brain to larynx is shorter. To trigger a verbal response, left-ear sound must travel to the right-brain, move down the neck into the chest and around the subclavian artery (that comes out of the aorta) and back up the recurrent nerve (part of the vagus network) to the larynx. To perceive the impetus for that vocal change, the left-brain must retrieve it from the right-brain that triggered it, which entails crossing nerves in the interhemispheric bridge (corpus callosum) to and from the right-brain.[27]

Tomatis tends to ignore the role of the left ear in normal behaviour because his emphasis was on making singers, especially, more strongly right-lateralized to improve their vocal pitch. However, those left-ear infusions of sound to the right-brain are important and their loss is evident in major behavioural disorders. They are intermittently lost to the left-brain when the hemispheres are poorly integrated, but they continue to influence the right-brain during its interval of dominance and incite audio and visual hallucinations.

Such gradations in control of the voice, both logical (left-brain) and emotional (right-brain), and variations in control of other kinds of behaviour are strongly evident in persons experiencing discontinuation syndrome. One subject in this study comments that her/his voice sounds robotic and that conversation has become impossible with speech reduced to a flat, monotonous, monosyllabic response. Those are symptoms of dysfunction in both ears, the right-ear's contribution of rational sentences and the left-ear's contribution of lilting variations in tone. His verbal deficits reflect the "bilateral distortions" in hearing Bérard recognized on the audiograms of dyslexic syndrome and autism patients.

States of Consciousness

Having discovered the role of the stapedius muscle in a complete spectrum of behaviours, from fully normal and intensely alert and creative thinking to the state of severe psychosis that most nearly resembles normal sleep and dreaming, I saw immediately the role of the normally flexible stapedius muscle in producing the changing levels of hemispheric integration we call "states of consciousness"—including high alertness, creativity, day-dreaming, prayer, meditation, sleep, dreams, coma, and so on. I also deduced the role of the ear in the acculturation and socialization of the maturational process that gradually trains the language of logic, belief, self-control, and deliberate action into the left hemisphere of the brain. Control of the ear muscles, although unconscious or semi-conscious, characterizes and enables "normal" behaviour, i.e., behaviour as it is socialized and acculturated. An inability to easily control the stapedius muscle will make it difficult for a person to change states of consciousness—to become alert, to go to sleep, to awaken, to distinguish visual impressions of reality from hallucinations, to integrate reason and imagination creatively, to distinguish reality from imagination, and in other ways to manage the usually differing and distinctive functions of the two cerebral hemispheres in an integrated fashion. Many such difficulties are evident among people withdrawing from SSRIs and psychoactive medications. Bérard cites other anomalies in audio profiles that explain other symptoms of these SSRI-withdrawal subjects, for example, the peaks at 1, 1.5, and 2 kHz that cause "allergic disorders such as asthma, hay fever, and eczema" and the hyperacusis peaks at 500 Hertz in both ears that characterize "aggressive personalities."[28] Tomatis often notes the close connection between the condition of the ear and the skin.

Categorizing the Data

The symptoms of 107 subjects were charted under the categories of dyslexia, chronic fatigue syndrome, depression, obsessive-compulsive disorder, bipolarity, and schizophrenia, symptoms that are familiar to me as being ear-related; under suicidal depression, which Bérard identified as ear-related; and under the few symptoms from Goldstein's "Checklist" that I had not observed directly. These categories include not only those parts of the inner ear commonly recognized for balance—the vestibule (I-A)—and hearing—the cochlea (I-B)—but also the stapedius muscle of the middle ear that Tomatis and Bérard treated for dyslexia and moderate CFS symptoms—the right stapedius muscle (II) —and that Bérard treated for suicidal depression—the stapedius muscle in either ear (III-A). A further middle-ear category contains severe CFS, schizoaffective, and schizophrenia symptoms—the right (and indirectly, the left)[29] stapedius muscle (III-B). A final category (Other) contains such symptoms as headaches that may not have been addressed with the Tomatis Method of ear stimulation, although Tomatis states that the audio-cervical loop also affects the lumbar region and sacrum,[30] which may be significant for headache and for feelings of pressure in the head and for piriformis muscle syndrome that causes sciatica. The "Checklist" also does not include headache or sensations of pressure in the head or sensations like electrical current running through the body ("zaps"), which frequently are symptoms of SSRI withdrawal. Tomatis noted that the stapedius muscle, an extensor, affects all the extensor muscles in the body just as the tensor tympani of the eardrum affects all the tensor muscles in the body[31] and I have placed all muscle symptoms, including erectile dysfunction and anorgasmia, in category II, accordingly.

Two Models of Neurological Intervention

To ameliorate the symptoms of his CFS patients Goldstein introduced chemicals into the trigeminal nerve,[32] the lower branch of which serves the muscle of the hammer (tensor tympani) of the middle ear, and the facial nerve, which serves the muscle of the stirrup (stapedius muscle) that regulates the inner ear.[33] Goldstein also contemplates altering the activity of the vagus nerve electronically.

> The vagus nerve is the other cranial nerve which modulates sensory input via the solitary tract in the brainstem. Devices to electrically stimulate the vagus have been implanted subcutaneously in the neck and have been helpful in treatment-resistant epilepsy. Such an approach should be useful in other neuroregulatory disorders.[34]

These are the nerve networks brought under control by the ear under the influence of high-frequency sound. A much better approach is to arouse normal function in a tiny muscle that responds to music. High-frequency music also has been used to heal severe epilepsy.[35] Tomatis notes that the audio-lumbar sacral loop and the audio-cervical loop function "under cochleo-vestibular supervision."[36] The vagus nerve that regulates the abdominal viscera also contributes to vocalization: "it extends a sensory nerve fibre onto the stirrup, which receives its motor activation [efferent] from the facial nerve."[37]

Goldstein identifies himself with the common medical paradigms of our period in history:

> My present treatments can also be viewed as resetting one or more neural networks but must be taken regularly to be effective, like insulin for diabetes. Almost all of these medications are quite safe if minimal precautions are taken. [38]

Anyone who has read Robert Whitaker's *Anatomy of an Epidemic*[39] will see Goldstein's simile of diabetes as a red flag. The well-intentioned doctor's rationalization is incorrect. It is the same rationalization that prevails among the pharmaceutical companies that extract chemicals from dyes, industrial disinfectants, and rocket fuel[40] to create SSRIs: "neuropharmacologic agents to produce a specific effect"[41] and persuade doctors to prescribe them for persons with damaged ears and central nervous systems already under duress. In fact, psychoactive medications are not at all like insulin for diabetes. They are not substances intrinsic to health or that restore a missing chemical the body is failing to produce. They are noxious chemicals that further damage normal processes to produce an outcome of symptom control. The outcomes of that treatment paradigm have been disastrous. Nevertheless, Goldstein proposes:

> Another technique to dynamically (and immediately) modulate the synaptic and intrinsic properties of neurons and reconfigure the networks they comprise is by externally altering trigeminal nerve input to the mesencephalic tract. . . . Immediate symptom alteration can be produced by certain kinds of eyedrops for the upper part, nasal dilators and adjustable upper teeth (maxillary) splints for the middle part . . . and sometimes lidocaine drops on the floor of the mouth (lower part).[42]

He also is open to cervical spinal surgery, "a decompression laminectomy," for some CFS patients, adapting a technique used for congenital spine deformities.

> Such a treatment is rational as I discussed in *Betrayal* since the upper cervical nerve roots comprise much of the mesencephalic tract [midbrain] of the trigeminal nerve, an important integrator of sensory input.[43]

How rational is it to chemically, electrically, and surgically alter nerves inadequately stimulated with the energy of sound if they might be restored through an intervention with high-frequency music that exercises the stapedius muscle to normal reactivity and thereby activates that entire neurological system?

Tomatis points out that the muscle of the jaw and of the first bone of hearing, the hammer, are controlled by the same nerve, a branch of the trigeminal nerve;[44] the stapedius muscle and those of the face near the seventh cranial nerve are innervated by the seventh cranial nerve (the facial nerve). The vagus nerve extends a sensory fibre onto the stapedius muscle, which is extremely important for the effect of that tiny muscle on the vast network of the vagus nerve.

> The vagus nerve, working in association with the glossopharyngeal nerve innervates the Eustachian tube and the pharynx. It also supplies sensory innervation to the larynx through the upper laryngeal nerve and provides motor function to the larynx through the lower laryngeal nerve, also called the recurrent nerve. This name is derived from the loop that this nerve makes around the right subclavian artery."[45]

That neurological network between the right ear and the larynx creates the audio-phono loop such that the voice can produce only the frequencies of sound likely to be heard by the ear (Tomatis Effect, 1957). The first phase of that cybernetic loop sends the vibrations of the singing voice as changes in air pressure caught by the outer ear (i.e., sound) and focused on the eardrum, through the middle and inner ear mechanisms through the brainstem to the upper temporal lobe, which I describe as the energy flow that maintains and enhances left-brain dominance.

Table 1 Withdrawal Symptoms of Persons Discontinuing One Antidepressant

Symptom Category	Number of Subjects	Symptoms Reported	Symptoms per Subject	Percentage of N=40	%age of Total Nr, Table 5
Vestibular Damage	12	15	1.25	30%	42%
Cochlear Damage	16	20	1.28	40%	35%
Vestibular and Cochlear Damage	8			20%	22%
Vestibular or Cochlear Damage	11			28%	27%
Moderate Middle-Ear Damage	31	193	3.0	78%	75%
Fatigue/ Insomnia	17			43%	41%
Psychosis	3	1	4.67	8%	18%
Suicide Attempts or Ideation	7			18%	26%
Severe Middle-Ear Damage	38	420	11.05	95%	97.6%
Other Symptoms	9	18	2.0	23%	29%
Total Residual Symptoms	40	560	14.0	100%	100%

The control of the voice is one of many cybernetic networks controlled by the ear and that are impaired by losses of left cerebral dominance when the stapedius muscle is dystonic—either genetically weak or damaged by some kind of assault. Similar symptoms occur when the cybernetic loop is interrupted (not in the ear but in the brain itself) by stroke.

Tomatis goes on to describe the right and left branches of the vagus nerve and their relationship to the bronchi, esophagus, coronary arteries, stomach; and to the aorta, most of the visceral organs, and the intestines. Nerves carry impulses of equal intensity at about the speed of sound.[46] It is impossible to avoid the relationship of the ear and the voice to the sympathetic and parasympathetic nervous systems.[47] These are the neurological systems Tomatis observed to respond positively to stimulation by high-frequency sound when he and his trained practitioners healed dyslexic syndrome, autism, and chronic fatigue syndrome. These are the neurological systems that normalize under stimulation by high-frequency sound to end difficulty swallowing, stomach upset, abdominal pain, gall bladder secretions, irritable bowel syndrome, and so on.

The chemical interventions Goldstein applies to the branches of the trigeminal nerve certainly affect the audio-phonal loop—which also is vital to left-brain dominance—but they achieve their effects in the same manner as SSRIs and other psychoactive drugs, that is, by temporarily altering a sense of malaise by disabling normal nerve pathways, not by restoring the neurology to normal activity. Goldstein mentions the "minimal precautions" of tests checking liver and kidney function for signs of damage from these chemical interventions, but he cannot speak to their risk *to the ear*, the role of which he is ignorant. As he is "resetting" the nerves linked to the ear, the possibility of damaging a network that might otherwise be restored appears to be high. The middle ear muscle is very sensitive to chemicals—even to caffeine[48]—and is easily damaged. Daniel's 0.0625 mg dosage of Risperdal had to be withdrawn for him to regain left-brain dominance (2006, 2008). His healing in 2016, which was entirely without medication, took half the time of his 2008 healing.[49]

Problems of the Inner Ear in CFS and SSRI Discontinuation

I have noted that my family's symptoms of CFS correlated very closely with those of Goldstein's patients. His "Checklist" has been widely applied for its correlation with the symptoms of other people diagnosed with CFS. From his study of the incidence of particular symptoms among his CFS patients, we learn that more than two-thirds of the more common symptoms fall into four categories:

1. **fatigue**: 95 per cent of CFS patients were fatigued;
2. **cognition changes**: 80 per cent suffered various forms of cognitive impairment, under which he lists attention deficit disorder (dyslexic syndrome);
3. **"psychological" problems**: 80 per cent had such symptoms as depression, panic attacks, and lability, including 1% with psychosis;
4. **"other nervous system problems"**: 75 per cent had nervous system symptoms, including disequilibrium and tinnitus.[50]

It has been very firmly established that many forms of cognitive impairment, including ADD, respond to music stimulation. We have seen, further, that many problems deemed "psychological," including suicidal depression, are *physiological* and respond to the physical stimulation of music. Although we may find it curious that among chronic fatigue syndrome patients 5% do not feel fatigue, we may also find it curious that 75% of CFS patients have symptoms all doctors should recognize as *ear-related*—balance problems, vertigo, nausea, and ringing in the ears (tinnitus). That incidence of inner ear problems should have alerted doctors immediately to the involvement of the ear in CFS. Tomatis explains that "The inner ear is a single entity. Any dysfunction in one of its parts leads to a more or less marked dysfunction of the entire system."[51] Yet, "Fitness of the muscles in the middle ear makes possible the optimal use of the inner ear. This requires an ongoing co-ordination of the muscles of the hammer and the stirrup.... The muscle of the stirrup regulates the inner ear. "[52] As Tomatis realized—and as the healing of three of our family member from CFS and Daniel's healing from dyslexia and schizophrenia demonstrate—the middle ear is the primary site of disorder of those three syndromes, as well as of the other behavioural patterns Daniel displayed during recovery from schizophrenia.

The stapedius muscle of the middle ear affects the movement of the stapes (stirrup), which impacts the endolymph of the inner ear through the membrane of the oval window to affect the vestibular canals and cochlea. Dystonia (myopathy) in the stapedius muscle deprives the vestibule of pressure waves in the endolymph that affect simultaneously the vestibular canals and the cochlea. An inactive stapedius muscle fails to protect the cochlea from loud sound and thus may allow cilia damage that results in tinnitus (although not all tinnitus arises in the ear per se). It is possible that such damage is the cause of misfiring of neurons that produces manic symptoms.[53] How the condition of the middle and inner ear may be an indicant of suicidal tendencies is discussed below.

The aggregate of symptoms of the cohort of people under study in withdrawal from Paxil and other SSRIs provides its own relative measures of severity. My personal experience and researched knowledge of similar symptoms in a broader but closely parallel range of suffering gives me insight into their symptoms. I rely on my sense of the position within a category of any one subject as compared with others in the Paxilprogress cohort and on my sense of the position of any subject in a category within my knowledge framework to categorize their data. Where a subject makes a vague reference to a set of symptoms without detailing the items in the set, I have supplied a *low* estimate of the discrete symptoms implied.

Symptoms of SSRI Discontinuation

1. Subjects and Collection of Data

One hundred and ten persons responded to an invitation from one member at the Internet forum Paxilprogress to volunteer information about their symptoms during and after withdrawal from Paxil or other psychoactive medications so that researchers could study the effects of these drugs. As is typical of the English Internet, subjects are primarily American, then Canadian and from the UK (England, Scotland, and Wales); others wrote from Australia and New Zealand, The Netherlands, France, Germany, and Sweden. Several did not provide country of origin information. The data used in this study were offered between 30 September 2006 and 1 September 2010 and were downloaded on the latter date.

2. Limitations of Data

The volunteers were not provided with a questionnaire or guidelines. Some assumed that information they had provided through posts on other threads in the forum discussions would be investigated further. Some provided links to their websites for additional information. Some of these leads were followed to augment the data provided.

3. Number of Subjects

Some subjects did not clarify whether their information applied to symptoms before, during, or after withdrawal. Some had not withdrawn completely from their medications. Some had withdrawn completely from Paxil[54] but continued to take other medication. Thus, the number of subjects who could be canvassed for information was reduced from the original number to N=107.

Table 2 Residual Symptoms of Persons Discontinuing One Antidepressant

Symptom Category	Number of Subjects	Symptoms Reported	Symptoms per Subject	Percentage of Total N
Vestibular Damage	39	73	1.9	46%
Cochlear Damage	33	44	1.3	39%
Vestibular and Cochlear Damage	18			21%
Vestibular or Cochlear Damage	46			62%
Moderate Middle-Ear Damage	57	192	3.37	68%
Fatigue/ Insomnia	25			30%
Psychosis	24	80	3.3	29%
Suicide Attemptsor Ideation	30			36%
Severe Middle-Ear Damage	76	945	12.8	91%
Other Symptoms	37	76	2.1	29%
Total Withdrawal Symptoms	84	1,368	12.79	100%

The number of subjects who reported using only paroxetine or whose reporting seemed to imply they had used only paroxetine (sold under the brand names of Paxil, Seroxat, Aropax) is n=41 or 38.3%. The number using just one SSRI is n=53 or 49.5%. Other SSRIs taken by subjects include sertraline hydrochloride [Zoloft, Selectra, Eleva], fluoxetine [Prozac, Sarafem], citalopram [Celexa, Cimpramil], and escitalopram [Cipralex, Lexapro]). Some other psychoactive drugs used by the subjects include the serotonin-norepinephrine reuptake inhibitor (SNRI) antidepressant venlafaxine (Effexor, Efexor), and benzodiazepines [clonazepam (Klonapin), Librium, and Valium], risperidone [Risperdal], and lithium). The number of subjects using other antipsychotics singly, consecutively, or in combination with paroxetine was n=67 or 62.6% of the total number of subjects (N=107).

As we are considering the cumulative effect of chemicals *on the ear* rather than the effects of reuptake inhibitors on the brain's chemistry, the variety of chemicals used by the subjects is not a barrier to analysis. The data collected are not sufficiently detailed to make comparisons between the various medications. The comparisons made in Tables 1 and 2 with the summary percentages of Tables 4 and 5 show that the symptoms produced by any one or by any combination of these medications are sufficiently similar to consider the cohort homogeneous, despite poly drugging. The overall effects of multiple drugs increase the percentage of subjects experiencing extreme fatigue and/or insomnia and "Other" symptoms. Poly drugging appears to suppress the number of symptoms of psychosis and to suppress more symptoms in the "Other" category.

4. Summary of Data

The data were gathered from responders' histories of withdrawal symptoms and of residual symptoms. The 2,212 symptoms of the subjects of the study have been listed in alphabetical order (to preserve anonymity) in Appendix B. Some subjects reported only withdrawal symptoms, only residual symptoms, or symptoms in both categories. Some listed side effects of the drug(s) that precipitated the decision to discontinue the SSRI, but these were excluded from the data. Differences in numbers of subjects reporting in any category are taken into account in those categories. The total number of subjects reporting only in the Withdrawal category is n=24. The total number of subjects reporting only in the Residual category is n=26. The total number of subjects reporting in both Withdrawal and Residual categories is n=57.

The amount of time just one antidepressant had been used (paroxetine or some other SSRI/SNRI) prior to withdrawal ranged from 0.3 years to 11 years. The period following withdrawal, i.e., the duration of residual symptoms, ranged from 4 months to 5 years. A comparison of subjects using just one antidepressant with the entire group of subjects using one or more medications indicates that in every major category, except insomnia and fatigue, those using only Paxil, Effexor, Lexapro, Klonapin, or Celexa make up a similar proportion of all subjects with residual symptoms. But among people who use *just one* antidepressant, fewer people experience vestibular symptoms (30% as compared with 42%), symptoms of psychosis (8% as compared to 18%), and suicidal tendencies (18% as compared to 26%). The relationship between the use of a single drug, a lower incidence of vestibular symptoms, and fewer people with suicidal intentions should be borne in mind.

5. Intervals of SSRI Use and Discontinuation

The period of time during which the subjects had used SSRI(s) varied from 8 days to 19 years (with a median of 6.0 years and a mean of 6.35 years); 90% of respondents had taken SSRIs for between 1 month and 13 years. The interval of time completely withdrawn from paroxitene and, in most cases, all other psychoactive medication at the time of reporting varied from 4 months to 8 years (with a median of 1.3 years and a mean of 1.21 years); 90% of respondents had been off SSRIs between 6 months and 5 years and still were struggling with symptoms.

Table 3 Intervals of Use and Non-Use of SSRIs

	Drug Use	Drug-Free
Range	<0.1 - 19 years	4 months - 8 years
Mean	6.35 years	1.21 years
Median	6.0 years	1.3 years
N =	75	68

The symptoms of withdrawal usually persist for such a long time following complete discontinuation from the medications that it is difficult to define the period of "withdrawal," which commonly means the interval of time between cessation and the return to normalcy or the baseline condition at the introduction of medication. In this study, the withdrawal period begins (1) when the drug is immediately and completely withdrawn ("cold turkey") and lasts for six months; but also (2) at the beginning of a period of negative titration, i.e., reduction in dosage.

The interval of negative titration can be from a few days or weeks to a year or longer. Those parameters should be recalled when noting the interval of use and the withdrawal symptoms, which for many subjects imply continuing SSRI use but at reduced dosages. Insofar as possible I have restricted Residual subjects to those who have ceased to take SSRIs for a minimum of 6 months.

Frequently, respondents mention withdrawal symptoms so severe they chose to reinstate the SSRI, resume a higher level of dosage, add another drug to the regimen, or switch to a different SSRI. The nature of SSRIs is to mask symptoms, which is why they were prescribed initially. They artificially raise serotonin (some, also norepinephrin) levels in the brain, which affects the serotonin–dopamine balance in the brain (and possibly in the ear).55 They do not replace a vital chemical missing from the body, although they frequently are misrepresented as doing so. Users often find that discontinuing the drug reveals the symptoms for which the drug originally was prescribed, but greatly exacerbated and with many more symptoms not experienced before taking the antidepressant medication, even when the period of use was as short as a few days or weeks (less than 2 months). Clearly, paroxetine and other psychoactive drugs damage some part of the body that (a) has already shown signs of dysfunction or damage and (b) has the capacity for further damage.

6. Severity of Symptoms

Some subjects provided much more extensive data than others, which may or may not reflect greater malaise. For example, a person with one residual symptom may have a very severe and life-threatening symptom despite the abating of other symptoms.[56] The naming of symptoms, both the number of items and their descriptions, may have been shaped by restraints affected by gender, cultural inclinations towards stoicism, individual variances in perception of pain, a developed capacity for introspection, memory of symptoms muted by the distance in time from an acute stage of suffering, or by other factors.

Table 4 Withdrawal Symptoms of Persons Discontinuing SSRIs

Symptom Category	Number of Subjects	Symptoms Reported	Symptoms per Subject	Percentage of Total N
Vestibular Dysfunction	39	73	1.9	46%
Cochlear Dysfunction	33	44	1.3	39%
Vestibular and Cochlear Dysfunction	18			21%
Vestibular or Cochlear Dysfunction	46			62%
Moderate Middle-Ear Damage	57	192	3.37	29%
Fatigue/ Insomnia	25			30%
Psychosis	24	80	3.3	18%
Suicide Attempts or Ideation	30			36%
Severe Middle-Ear Damage	76	945	12.8	91%
Other Symptoms	37	76	2.1	29%
Total Residual Symptoms	84	1,368	12.79	100%

While the data were not elicited in an organized way by such means as a questionnaire or by the provision of categories, it is likely that respondents read the medical histories other subjects had posted. Most respondents had extensive knowledge of other persons' withdrawal symptoms from having frequented the forum for months or years. A person's self-reporting could be shaped to some extent by that commonality, comparing, and sharing. Many conducted research into their symptoms and read Internet articles on side effects of their medications that may have framed their perception of their symptoms.

All respondents were or had been under the care of at least one physician and had identified, clarified, and explored the origins of their symptoms, although not necessarily to their satisfaction as patients, in a physician's office, emergency room, hospital, or some combination of those venues. The naming of their symptoms was in no instance vague or naïve. However, they tend not to imply how severe they consider their symptoms to be except in a few cases.

Table 5 Residual Symptoms of Persons Discontinuing SSRIs

Symptom Category	Number of Subjects	Symptoms Reported	Symptoms per Subject	Percentage of Total N
Vestibular Dysfunction	35	41	1.17	42%
Cochlear Dysfunction	29	37	1.28	35%
Vestibular and Cochlear Dysfunction	18			22%
Vestibular or Cochlear Dysfunction	22			27%
Moderate Middle-Ear Damage	62	193	3.11	75%
Fatigue/ Insomnia	34			41%
Psychosis	15	44	2.93	18%
Suicide Attempts or Ideation	21			26%
Severe Middle-Ear Damage	81	780	9.63	97.6%
Other Symptoms	24	41	1.7	29%
Total Residual Symptoms	83	1,095	13.2	100%

Some indication of the severity of symptoms is that many of the respondents have been too ill to work at their former means of employment or to continue with their education or to care for themselves, their homes, or their families.

7. Changes in Severity of Symptoms during and after Discontinuation

The type of symptoms during withdrawal and remaining as residual symptoms after 6 months to a year do not appear to differ according to the length of time the drug or drugs were taken. It is difficult to assess the severity of symptoms in individual cases for the reasons described above. However, in the aggregate, the pattern of residual symptoms seems almost as serious as those during withdrawal. For example, the incidence of suicide attempts and suicidal ideation in withdrawal is 35.7% of all subjects and among those who had been off their medication for at least six months it is 26.3%. The total number of serious symptoms reported (category IIIB),

Table 6 Suicide Ideation/Attempts and Symptoms of Ear Damage of Subjects in Withdrawal from SSRIs

Suicide Attempts or Ideation	Vestibule Only N=39	Cochlea Only N=33	Inner Ear (Both V+C N=18	Middle Ear, Moderate N=57	Middle Ear, Severe N=76
Suicidal Subjects in Withdrawal n=30	6	3	9	23	30
% of All Subjects with Withdrawal Symptoms (N)	15%	1%	50%	40%	39%
Suicidal Subjects with Psychosis n=15	2 13% of n	2 13% of n	6 40%	10 67% of n	15 100% of n

excluding suicide, during withdrawal is 945 compared with a greater number of serious residual symptoms, excluding suicide, which is 1,095. The total in both categories, 2,040, differs from the total in Appendix B, 2,212, because estimates of discrete symptoms were made of composite responses, such as "ADD," which includes numerous symptoms. The data used for analysis are slightly less than those reported.

Some subjects describe the gradual moderation of some of their symptoms during and following discontinuation. Five respondents of 107 claim complete or full recovery, although one of these notes a remaining quite severe symptom and another was discontinuing another medication.[55]

By contrast to "full recovery," most of the respondents have improved to some extent but 103 of 107 persons have continuing symptoms, many of them severe, and some even 6 years after the cessation of medication.

Discussion of Data

The Inner Ear—Vestibule I-A; Cochlea I-B

The distribution of all subjects' symptoms according to their origins in the middle and inner ear is shown in Tables 4 and 5. The first two sub-categories pertain to the inner ear: the vestibular labyrinth (category I-A) and the cochlea (category I-B). The occurrence of such symptoms as vertigo and disequilibrium or tinnitus pertaining to these parts of the inner ear is widely recognized.

Table 7 Suicide Ideation/Attempts and Symptoms of Ear Damage of Subjects with Residual Symptoms

	Vestibule Only N=35	Cochlea Only N=29	Inner Ear (Both V+C) N=20	Middle Ear, Moderate N=62	Middle Ear, Severe N=81
Suicidal Subjects with Residual Symptoms n=21	3	3	4	21	21
% of All Subjects with Residual Symptoms (N)	1%	20%	26%	26%	26%
Suicidal Subjects with Psychosis n=9	1 5% of n	1 5% of n	3 14% of n	6 67% of n	9 100% of n

Significantly, 62% of persons in withdrawal and 63% of persons with residual symptoms report problems with the inner ear, either the vestibule (46% and 42%) or cochlea (39% and 35%) or both (21% and 22%). As mentioned above in reference to chronic fatigue syndrome, these sorts of inner ear problems, which 75% of Goldstein's CFS patients also have, coincide with middle ear stapedius muscle dysfunction because "[T]he muscle of the stirrup regulates the inner ear."57 Thus, the sorts of inner-ear problems 62–63% of SSRI users report also could be related to stapedius muscle dysfunction in the middle ear. Inner-ear symptoms, especially when vestibular and cochlear symptoms occur together, have a bearing on suicide, as discussed below.

The Middle Ear—Moderate Damage II

Symptoms recognized by Tomatis as pertaining to the middle ear, and that characterize moderate CFS and dyslexic syndrome, are classed in category II. Symptoms of moderate middle ear damage are displayed by 68% of subjects in withdrawal from SSRIs and by 75% of subjects with residual symptoms. All 30 subjects in withdrawal who have suicidal tendencies have symptoms of moderate damage to the middle ear.

The Middle Ear—Severe Damage III-B

Symptoms of severe chronic fatigue syndrome, bipolar II and I (without and with hallucinations, respectively), psychosis, autism, and other severely aberrant behaviour patterns are classed in category III-B. Subjects in withdrawal report an average of 12.79 severe symptoms each. Those who have ceased to take SSRIs for at least 5 months or more report an average of 9.63 severe symptoms. A high preponderance of respondents in withdrawal—98%— have symptoms of severe damage to the middle ear and among those with residual symptoms fully 91% still have symptoms of severe middle ear damage. All 21 subjects who ceased taking SSRIs and who have suicidal tendencies also have symptoms of severe damage to the middle ear.

Fatigue/Insomnia

The exceptional fatigue and severe insomnia experienced by many subjects was recorded separately as a sub-category of III-B. The inability of the stapedius muscle to process sufficient high-frequency sound to the left hemisphere to keep it dominant can produce hallucinations.58 Fatigue also is the result of insufficient energy reaching the left hemisphere of the brain, but not so little as to unhinge dominance. The inability of the stapedius muscle to relax so that the hemispheres integrate at a lower level to induce sleep is a cause of insomnia. Among SSRI discontinuation subjects with withdrawal and residual symptoms, 30% and 41%, respectively, experience incapacitating losses of energy and/or severely compromised ability to sleep.

Psychosis

Such symptoms as audio hallucinations that involve the inner ear have not been assigned to the inner ear categories but are included in the sub-categories of psychosis because they entail severe damage to the middle ear (category III-B). Among withdrawal subjects 29% have experienced an average of 3.3 symptoms of psychosis and 18% of residual subjects have experienced an average of 2.93 symptoms of psychosis. Most of these individuals report having visual hallucinations, which are a clear symptom of intervals of hemispheric integration disorder, i.e., of non-dominance of the cerebral hemispheres.59 Subjects do not clarify whether their psychosis occurred in brief episodes (bipolar I) or prevalently (as in autism, schizophrenia, etc.).

Both during withdrawal and residually, 29% of respondents have symptoms categorized as Other, including headache, feelings of pressure in the head, "zaps" (which may indicate shifting of dominance between hemispheres, which would remove those to category III-B), and other symptoms of uncertain origin.

Suicide Ideation and Attempts— III-A

Suicide ideation and attempts at suicide have been shown by Bérard to arise in a malfunction of either ear, although predominately of the left ear. The severity of the symptom, as well as its ambivalent etiology (as regards cerebral dominance), has earned this symptom a separate category (III-A).

Although ingesting a drug that damages the right ear presumably also damages the left ear, the prevalence of right-handedness among humans (>80% of persons) and of left-brain dominance creates certain expectations and norms in the population generally. No information about laterality was asked for or offered by these subjects. Bérard does not provide that information about his suicidal patients, unfortunately. Schizophrenia is associated with left-handedness, which is not surprising because the condition of left-handedness tends to reduce the strength of left-brain dominance.60 It is possible that those of Bérard's patients who were suicidally depressed and were relieved of their depression through stimulation of their left ears also were left-lateralized. Normalizing the dominant ear of a left-handed person may restore left-brain dominance, including rational self-control that could obviate suicide. The data here indicate damage to the middle ear in 36% of withdrawal subjects (Table 2) and in 26% of subjects with residual symptoms (Table 3). Table 4 shows how the suicide ideation and attempts of those people correlate with the information given that pertains to their inner ears. As the respondents were not prompted for symptoms concerning their inner ears, such symptoms may be under-reported.

While an indication of an inner ear problem is not proof that a person discontinuing SSRIs will contemplate or attempt suicide, more than half of those who do have suicidal tendencies report symptoms of dysfunction in the vestibule, cochlea, or both parts of the inner ear (6+3+9 = 18 out of 30 in withdrawal and 3+3+4 = 10 out of 21 with residual symptoms). Among suicidal subjects reporting only one inner ear symptom, twice as many reported vestibular symptoms as those who reported cochlear symptoms.

Notably, 100% of persons with suicide ideation or who attempted suicide during withdrawal had symptoms of severe middle ear damage (category III-B), and all but 7 (23 of 30) had symptoms of moderate ear damage (category II) as well. Of those with residual symptoms that included suicidal and/or homicidal tendencies 100% had symptoms of severe middle ear damage (category III-B), and all but 5 had symptoms of moderate ear damage (category II) as well.

Whether in withdrawal or when dealing with residual symptoms, more than 50% of the people at either stage of discontinuation who show moderate symptoms of middle ear damage contemplate or attempt suicide.

Of all subjects who show signs of severe middle ear damage, 39% in withdrawal and 26% with residual symptoms contemplate or attempt suicide or homicide (2 cases).61 Among all subjects with suicidal tendencies, whether in withdrawal or with residual symptoms, 68% have symptoms of moderate middle ear damage and 100% have symptoms of severe middle ear damage. Too few subjects clarified their suicidal ideation as compared with actual attempts at suicide or homicide to allow separation of those categories. One person had made five attempts on his or her life. One person mentioned only suicide ideation but also that her husband had kept constant watch over her. Another subject said that his gun and car keys had been taken from him and he lived with his parents under their supervision. Most stated emphatically that they had never contemplated or attempted suicide prior to taking SSRI medication. Obsessive thoughts on any subject are a result of low hemispheric integration, which tends also to remove the capacity for deliberate action. That is one of the few things that may be positive about middle ear damage: at least one failsafe control is built into a non-dominant hemisphere and that is its (usually) limited ability to act deliberately. However, highly trained behaviour, such as produced in the military, can result in homicidal or suicidal behaviour during losses of left-hemisphere dominance.

The degree of damage to the middle ear stapedius muscle is a continuum from none to irretrievable paralysis that can be measured in Hertz and that produces a spectrum of particular behaviours. These data have been categorized in artificially segregated blocks that obscure fine gradations and that do not account for all possible complications. If the suicidal subjects' ears could be precisely tested for the ability to perceive high-frequency sound it is possible that each ear would show a different profile for audio-processing of particular frequencies of sound, as Bérard found for people with autism and dyslexic syndrome. The differences between the two ears' ability to process sound may exacerbate episodes of alternating dominance: an inter-hemispheric battle that undercuts the ability to preserve the unified, left-brained self-concept,62 which is the definition of schizophrenia but also manifests during the outbursts that punctuate bipolar behaviour or even dyslexic syndrome. Multiple ear problems, such as the masking in Daniel's right ear that probably made him more reliant on his left ear for months, can prejudice the recovery of left-brain dominance. Binaural treatments also reinforce the left ear and undermine the effort to establish dominance in the right ear. Such complications need to be taken into consideration when analyzing the effects on the brain and on behaviour of various kinds of damage to the ears and of other music therapies. Persons who have experienced total hearing loss, for example, due to the surgical removal of a tumour that destroys the auditory and vestibular nerves, and then is given antidepressants for the brain

Suicide Ideation and Attempts— III-A

Suicide ideation and attempts at suicide have been shown by Bérard to arise in a malfunction of either ear, although predominately of the left ear. The severity of the symptom, as well as its ambivalent etiology (as regards cerebral dominance), has earned this symptom a separate category (III-A).

Although ingesting a drug that damages the right ear presumably also damages the left ear, the prevalence of right-handedness among humans (>80% of persons) and of left-brain dominance creates certain expectations and norms in the population generally. No information about laterality was asked for or offered by these subjects. Bérard does not provide that information about his suicidal patients, unfortunately. Schizophrenia is associated with left-handedness, which is not surprising because the condition of left-handedness tends to reduce the strength of left-brain dominance.60 It is possible that those of Bérard's patients who were suicidally depressed and were relieved of their depression through stimulation of their left ears also were left-lateralized. Normalizing the dominant ear of a left-handed person may restore left-brain dominance, including rational self-control that could obviate suicide. The data here indicate damage to the middle ear in 36% of withdrawal subjects (Table 2) and in 26% of subjects with residual symptoms (Table 3). Table 4 shows how the suicide ideation and attempts of those people correlate with the information given that pertains to their inner ears. As the respondents were not prompted for symptoms concerning their inner ears, such symptoms may be under-reported.

While an indication of an inner ear problem is not proof that a person discontinuing SSRIs will contemplate or attempt suicide, more than half of those who do have suicidal tendencies report symptoms of dysfunction in the vestibule, cochlea, or both parts of the inner ear (6+3+9 = 18 out of 30 in withdrawal and 3+3+4 = 10 out of 21 with residual symptoms). Among suicidal subjects reporting only one inner ear symptom, twice as many reported vestibular symptoms as those who reported cochlear symptoms.

Notably, 100% of persons with suicide ideation or who attempted suicide during withdrawal had symptoms of severe middle ear damage (category III-B), and all but 7 (23 of 30) had symptoms of moderate ear damage (category II) as well. Of those with residual symptoms that included suicidal and/or homicidal tendencies 100% had symptoms of severe middle ear damage (category III-B), and all but 5 had symptoms of moderate ear damage (category II) as well.

Whether in withdrawal or when dealing with residual symptoms, more than 50% of the people at either stage of discontinuation who show moderate symptoms of middle ear damage contemplate or attempt suicide.

Of all subjects who show signs of severe middle ear damage, 39% in withdrawal and 26% with residual symptoms contemplate or attempt suicide or homicide (2 cases).61 Among all subjects with suicidal tendencies, whether in withdrawal or with residual symptoms, 68% have symptoms of moderate middle ear damage and 100% have symptoms of severe middle ear damage. Too few subjects clarified their suicidal ideation as compared with actual attempts at suicide or homicide to allow separation of those categories. One person had made five attempts on his or her life. One person mentioned only suicide ideation but also that her husband had kept constant watch over her. Another subject said that his gun and car keys had been taken from him and he lived with his parents under their supervision. Most stated emphatically that they had never contemplated or attempted suicide prior to taking SSRI medication. Obsessive thoughts on any subject are a result of low hemispheric integration, which tends also to remove the capacity for deliberate action. That is one of the few things that may be positive about middle ear damage: at least one failsafe control is built into a non-dominant hemisphere and that is its (usually) limited ability to act deliberately. However, highly trained behaviour, such as produced in the military, can result in homicidal or suicidal behaviour during losses of left-hemisphere dominance.

The degree of damage to the middle ear stapedius muscle is a continuum from none to irretrievable paralysis that can be measured in Hertz and that produces a spectrum of particular behaviours. These data have been categorized in artificially segregated blocks that obscure fine gradations and that do not account for all possible complications. If the suicidal subjects' ears could be precisely tested for the ability to perceive high-frequency sound it is possible that each ear would show a different profile for audio-processing of particular frequencies of sound, as Bérard found for people with autism and dyslexic syndrome. The differences between the two ears' ability to process sound may exacerbate episodes of alternating dominance: an inter-hemispheric battle that undercuts the ability to preserve the unified, left-brained self-concept,62 which is the definition of schizophrenia but also manifests during the outbursts that punctuate bipolar behaviour or even dyslexic syndrome. Multiple ear problems, such as the masking in Daniel's right ear that probably made him more reliant on his left ear for months, can prejudice the recovery of left-brain dominance. Binaural treatments also reinforce the left ear and undermine the effort to establish dominance in the right ear. Such complications need to be taken into consideration when analyzing the effects on the brain and on behaviour of various kinds of damage to the ears and of other music therapies. Persons who have experienced total hearing loss, for example, due to the surgical removal of a tumour that destroyes the auditory and vestibular nerves, and then is given antidepressants for the brain's reaction to losing half its sound supply, will suffer the further consequences of that drug assault on the stapedius muscle of the remaining viable ear.

These data indicate that middle-ear problems among suicidal subjects are not entirely at the most severe end of the spectrum, as would be the case if they were schizophrenic or autistic. However, the reporting is not clearly delineated in time; a subject may have passed through a suicidal phase while recovering from a psychotic phase. Another subject (ex-military) ideates murdering his wife and "other family member," his dog, and has been doing so for two years without acting on his obsessive thoughts. Some of these persons do list a few symptoms of psychosis (16 of 30 subjects in withdrawal; 9 of 21 subjects with residual symptoms). The presence of psychotic symptoms does not imply that the person is suicidal. It is possible that the involvement of one damaged ear (the "2–8" profile) in a situation of non-dominance propels suicidal tendencies. This is an important area for further research.

Some suicidal subjects experiencing psychosis (7 in withdrawal and 5 with residual symptoms) lack moderate middle-ear symptoms but have quite severe middle-ear symptoms.

Thus, their suicidal behaviour may have appeared in brief episodes, placing them in the bipolar I sector of the spectrum during that phase of their recovery; or, if their symptoms were more consistent, in the schizophrenic sector of the spectrum during that phase of their recovery. It would be very helpful to know whether suicidal behaviours were in any rational sense deliberate (e.g., to escape suffering, as one relative of a suicide surmised) or were the spontaneous, uncontrolled sort of dangerous behaviour that can occur during the right-brain phase of the cycle of schizophrenia. However, the data lack sufficient detail for that analysis.

Summary

The symptoms of moderate and severe middle ear damage among subjects in SSRI withdrawal (192+945=1,137) is modestly greater (164 symptoms or 14% of that total) than symptoms of moderate and severe middle ear damage among subjects experiencing residual symptoms (193+780=973). The average number of middle-ear symptoms per subject in withdrawal is 14; the average number of residual middle-ear symptoms for a subject is 11.7. The symptoms of persons who have reduced and withdrawn SSRI medication are consistent with established knowledge of the effects on behaviour of damage to the middle and inner ear.

Most of the presenting problems for which SSRIs were prescribed for these subjects have fewer than half a dozen symptoms: sorrow in common life circumstances such as the death of a parent or a miscarriage; anxiety over work, a failed relationship or broken marriage; insomnia; or mild depression. A person with middle- and inner-ear dysfunction cannot manage emotional reactions to common sources of emotional distress—or physical pain—within the socialized range of normal behaviour. While other presenting problems were in some cases more severe, such as major depression or possibly autism, evidence indicates that these acute conditions, also, are ear-related and might be treatable with high-frequency sound. Bérard's finding that specific audio profiles are associated with suicidal depression, milder depression, and aggression, affirm the author's hypothesis that not only mental illnesses, but such conditions as discontinuation syndrome, describe a spectrum that might be calibrated to specific audio-processing deficits identifiable on an audiogram in either the right, the left, or both ears.

Notes

[1] Andrews, P.W., Thomson, J.A. Jr., Amstadter, A., and Neale, M.C. (2012) "Primum non nocere: an evolutionary analysis of whether antidepressants do more harm than good", *Frontiers of Psychology* 3: 117. doi: 10.3389/fpsyg.
2012.00117 Received: 31 December 2011; Accepted: 30 March 2012; Published online: 24 April 2012.

[2] Robert Whitaker elaborates on the damage done to people by SSRIs and other psychoactive medications in *Anatomy of an Epidemic: Magic Bullets, Psychiatric Drugs, and the Astonishing Rise of Mental Illness in America* (New York: Crown, 2010). See, also, the forum http://survivingantidepressants.org and https://fiddaman.blogspot.com.

[3] Gregory, Richard L., *Mind in Science*. Harmondsworth, Middlesex: Penguin Books, 1981, 203.

[4] Although Dr. Bérard mentions having treated people successfully who had been subjected to psychoactive medications, see Guy Bérard, *Hearing Equals Behavior* (New Canaan, Conn.: Keats, 1993), 152, 156. References to Bérard are from this edition although the 2011 edition updated and expanded by Sally Brockett gives a few new details of his thinking and research.

[5] Daniel's ears probably were injured in utero through protracted exposure over several days to very loud sound from an electric table saw I used to build a wall unit of eight bookcases.

[6] In the late 1980s, a small group of physicians studied a syndrome of symptoms known by several names: myalgic encephalitis (ME), chronic fatigue immune deficiency syndrome (CFIDS), chronic fatigue syndrome (CFS), and fibromyalgia. The epidemic of these symptoms also was known by such

unhelpful terms as "the yuppie flu." Like other similar epidemics of those symptoms, the one that began in 1988 followed an outbreak of influenza, referred to at that time as "the Asian flu."

7 One of these physicians, Jay Goldstein, studied the incidence of his patients' symptoms. That summary, which he called "Looking in the Mirror, A Checklist of ME/CFIDS Symptoms," became widely used by other doctors in preliminary consultation with patients who might be suffering from ME/CFIDS. Goldstein wrote two books about his efforts to understand CFS and his attempts to alleviate the suffering of his patients.

8 My very brief use of Zoloft produced an alarming mental event that I recognize now as having reduced the integration of the two halves of my brain: I had rapid total recall of streams of words in my memory with only a dim awareness that I should know what they mean, but with no ability to order them or to control the eruption of long-term memory.

9 Dr. Alfred Tomatis (1920–2001), a French otolaryngologist, who conducted research adjacent to his medical and surgical practice as an otolaryngologist, was decorated by the French government for his contributions to science.

10 I describe Daniel's healing from dyslexic syndrome by the Tomatis Method in my book *Listening for the Light*, chs. 3–5.

11 Alfred A. Tomatis, *The Ear and the Voice* (Lanham, Md.: Scarecrow Press, 2005 [1987]), 9.

12 Ibid.

13 Bérard, a former colleague of Tomatis, founded the Bérard's Audio Integration Training (AIT) network of practitioners applying electronically filtered high-frequency sound binaurally.

14 I assume that those persons attempting withdrawal of SSRIs who become fully schizophrenic are unable to participate coherently in an online forum or that they have been subdued with medication that makes such communication difficult. Details of various assaults on Daniel's ears that produced psychosis are in *Listening for the Light*.

15 Robert Whitaker describes the historical trends in psychiatry that led to the use of psychoactive medications that alter neurology without healing it, but, in fact, exacerbate conditions that might heal spontaneously or through more benign types of treatment. *The Anatomy of an Epidemic* (New York: Crown, 2010), 67–85.

16 Ibid., 156–9, 194–6, passim. Robert Whitaker elaborates on chronicity of mental states and increased psychosis in schizophrenia caused by psychoactive medications.

17 I describe the origins of concepts of "good" and "evil" and of "heaven" and "hell" as attached to the left and right cerebral hemispheres, respectively, when analyzing Daniel's language during schizophrenia in *Listening for the Light*, chs. 12 and 13.

18 And of less severe depression to a "1.5—8" profile or even less severe depression to a "1—8" profile: Guy Bérard, *Hearing Equals Behavior*, 45–51, 140-62. Here, and in many other places in Bérard's and Tomatis's theorizing about the ear and behaviour, their lack of understanding of hemispheric dominance and losses thereof limits their contextual descriptions of the specific phenomena observed. Note that in Bérard and Brockett, *Hearing Equals Behavior: Updated and Expanded* (Manchester Center, Vt., 2011), ch. 14, the various schema offered for explaining the means by which AIT effects change and cure all are partial and some are speculative and none recognize the mechanism for lateralization through right-ear-driven cerebral dominance.

19 Bérard's book title asserts "*Hearing Equals Behavior*," but he offers no examples of acute adult mental illness other than suicidal depression. He warns against using his method for schizophrenia. His writing draws on more than 8,000 patient profiles of which his 235 suicidally depressed patients, of course, were only a relatively few. See, also, the research summary at Spot, Centro Terapéutico y de Investigación at: <www.spotcentre.es/investigaciones.html>. (15 November 2010)

Dr. Guy Bérard, at his clinic in France reported in 2005, the following information:

Depression, suicidal—233 cases treated over 235 cases reviewed:

217 healings from the first treatment, i.e., 93%

11 cures after two or three treatments, i.e., 4.7%

4 failures after treatment

1 case of failure to start treatment.
The two cases examined but not treated ended in suicide at 18 months. From the book "Hearing Rehabilitation" [the Spanish translation of *Hearing Equals Behavior*] by Dr. Guy Bérard as quoted at: <www.spotcentre.es/investigaciones.html>.

[20] Tomatis tried unsuccessfully to convince his Canadian associates that mental illness originates in the ear. *The Conscious Ear, My Life of Transformation through Listening* (Barrytown, NY, and Sound Listening and Learning Center, Phoenix, Ariz.: Station Hill Press, 1991) 200–1. Tomatis's theorizing was tied to a Freudian model with which he was uncomfortable, ibid., 216–19, that assumed a capacity for volition in the autistic child and, by extension, to all persons with audio-processing deficits. Daniel's example indicates the reverse to be the case: that auditory failure undermines the capacity for volitional acts and for learning self-control of all kinds.

[21] Tallman, *Listening for the Light*, ch. 11.

[22] Richard L. Gregory, *Oxford Companion to the Mind* (London: Oxford University Press, 1987), 305.

[23] And perhaps most famously in the artist Vincent van Gogh.

[24] Tallman, *Listening for the Light*, ch. 13.

[25] Tomatis, *The Conscious Ear*, 201, 208.

[26] Tomatis, *The Ear and the Voice*, 65.

[27] Creating the inter-hemispheric transfer time (IHTT). See also Tomatis, *The Ear and the Voice*, 75.

[28] Bérard, *Hearing Equals Behavior*, 42, 48–9.

[29] Although Daniel was right-handed and recovered through stimulation to his right ear, persons with schizophrenia and schizoaffective disorders more frequently are left-handed, indicating an underlying problem in the right ear that has forced the left ear to become the dominant, listening ear. However, when the right ear is less sensitive at any frequency, the more sensitive left ear will take over at that frequency. When the left ear, overall, is providing a fragile predominance of energy to the left brain and then becomes damaged, that dominance fails. (see, for example, Mason, P., Rimmer, M., Richman, A., Garg G., Johnson J,. and Mottram P. 2008. "Middle-ear disease and schizophrenia: case-control study," *British Journal of Psychiatry*, 193: 192—6.).

[30] Tomatis, *The Ear and the Voice*, 76.

[31] Tomatis, *The Ear and the Voice*, 59.

[32] Jay A. Goldstein, M.D., "Treatment Protocol for Physicians," <www.ncf-net.org/forum/jay.htm>. Accessed 14 April 2009.

[33] Tomatis, *The Ear and the Voice*, 52.

[34] Goldstein, "Treatment Protocol."

[35] Regarding "static epilepticus, the worst form of epilepsy", Dr. Jeff Bradstreet, "Foreword," in Annabel Stehi, ed., *Sound of Falling Snow: Stories of Recovery from Autism and Related Conditions* (New York: Beaufort Books with the Georgiana Institute, 2004), x. See also Tomatis's treatment of epilepsy with his Electronic Ear, Tomatis, *The Conscious Ear*, 189. I also have had the opportunity of observing seizures following stroke and following prescribed or illicit drug use, all of which leads me to believe ear function is involved in such brief episodes of psychosis.

[36] Tomatis, *The Ear and the Voice*, 76.

[37] Tomatis, *The Ear and the Voice*, 62–3.

[38] Goldstein, "Treatment Protocol."

[39] Whitaker, *Anatomy of an Epidemic*, 47.

[40] Whitaker, Anatomy of an Epidemic, Ch. 4.

[41] E.H. Parsons, psychiatrist, at a 1955 meeting in Philadelphia; quoted in Whitaker, *Anatomy of an Epidemic*, 51.

[42] Goldstein. "Treatment Protocol."

[43] Ibid. (my emphasis).

[44] Tomatis, *The Ear and the Voice*, 60.

[45] Ibid., 62.

[46] "The pulses of equal intensity and the 'all-or-none' response of nerve[s] were described by E.D. Adrian in *The Basis of Sensation* in 1928." Richard L. Gregory, *Mind in Science* (Harmondsworth, Middlesex: Penguin Books, 1981), 203–5.
[47] Tomatis, *The Ear and the Voice*, 62–3.
[48] Matthew C. Holley, "Application of new biological approaches to stimulate sensory repair and protection," *British Medical Bulletin* 63 (2002): 16–19. Holley discusses a number of agents, including antibiotics, that attack the cilia and notes several strategies being developed for stimulating cell regrowth. He suggests agents, including aspirin, that may protect the cilia prior to undergoing potentially damaging drug interventions. It is not clear whether sound damage might also be deflected by this means.
[49] Laurna Tallman, *Awakening Normal: How Focused Listening to Music Heals Mental Illness, A Neurological Paradigm* (Marmora, Ont.: Northern Light Books, 2018), p.60.
[50] Goldstein, "ME/CFIDS Checklist."
[51] Tomatis, *The Ear and the Voice*, 52.
[52] Ibid.
[53] Willot, J.F., and S.M. Lu. 1982. "Noise-induced hearing loss can alter neural coding and increase excitability in the central nervous system of mice," *Science* 216, 4552 (1982): 1331. At: <www.sciencemag.org/cgi/content/abstract/216/4552/ 1331>.
[54] Paxil (paroxetine) is sold under various brand names in different countries by GlaxoSmithKline, including Paxil (US), Seroxat (UK), Aropax (Australia).
[55] A study of dopamine within the cochlear system itself is G. Halmos, B. Lendvai, et al., "Simultaneous measurement of glutamate and dopamine release from isolated guinea pig cochlea," *Neurochem Int* 40, 3 (2002): 243–8.
[56] Three years post-Paxil use: Dysautonomia, unable to maintain blood pressure through a normal diet; requires large amounts of fluid and salt.
[57] Tomatis, *The Ear and the Voice*, 52.
[58] Tallman, *Hemispheric Integration and the Ears: A Scientific and Inclusive Paradigm of Human Behaviour including the Mild and Severe Forms of Mental Illness* (Marmora, Ont.: Northern Light Books, 2011), 12.
[59] Ibid., 11.
[60] Tomatis alludes to problems produced by left-laterality in the ears, *The Ear and the Voice*, 22; *The Conscious Ear*, ch. 10. I describe the greater fragility of left-brain dominance in left-lateralization in *Hemispheric Integration and the Ears*, 28, and in *Listening for the Light*, 90–1, and 302–3.
[61] Psychologist Dr. Ann Tracy reports on mass homicides and other killings associated with the use of SSRIs at: http://www.ssristories.com/index.php (accessed again 14 December 2012, the day of another such massacre in Newtown, Connecticut).
[62] Tallman, *Listening for the Light*, ch. 12.

Appendix C Withdrawal and Residual Symptoms of Persons Discontinuing Paxil and/or Other Psychoactive Medications

insomnia IIIB
emotional numbness IIIB
anxiety, episodes of intense IIIB
depression unrelated to environmental events IIIB
concentration problems II
memory problems IIIB
cognitive problems II
hypersensitivity to medications IIIB
hypersensitivity to other substances IIIB
anorgasmia II
mood swings, wide range of IIIB
suicidal IIIA
sensitive to noise I-B
agoraphobia II
concentrate, unable to II
fears, unreal IIIB
nightmares,. horrific IIIB
self harm IIIB
headache, chronic O
light sensitivity, extreme IIIB
depersonalization IIIB
anxiety IIIB
confusion II
memory problems II
visual lag IIIB
dizziness IA
sound in my right ear, clicking IB
suicide, attempted III-A
visual disturbances ("whoosh effect") IIIB
zaps, electrical sensations IIIB
memory loss, short term (documented with standardized testing) IIIB
agorophobia IIIB
hallucinations, visual
hallucinations, auditory) IIIB
insomnia IIIB
anxiety, increased IIB
concentration, decreased II
memory, decreased retention of information from the written word. II
suicidal IIIA
dysautonomia, unable to maintain his blood pressure through a normal diet. Now requires large amounts of fluid and salt. O
anxiety, extreme IIIB
panic and fear IIIB
weepiness, bouts of IIIB
head pressure O
chest and back, terrible stress feeling in O
"morning anxiety" I never had this problem before. IIIB
Anxiety, persistent high IIIB
heart palpitations (which did not correspond to the times of highest anxiety) IIIB
arthritic pain in knees, extreme IIIB
cognitive problems
concentrate on reading or writing inability to II
memory loss IIIB;
cognitive, difficulty with abstract thought) II
blistering and inflammation inside mouth O.
sleepiness, perpetual IIIB
tearfulness IIIB
suicidal depression IIIA

depersonalization IIIB
sexual function, altered (alternating low libido and hypersexuality, severe premature ejaculation) IIIB
breathlessness O
rages IIIB
hair thinning / parts of hair and beard suddenly turning snow-white O
tinnitus IB
mental reactions, slow, and sluggishness II
swollen and extraordinarily prominent veins in arms and legs O
stress, inability to cope with II
derealisation and IIIB
depersonalisation IIIB
rage
mood swings, IIIB
suicide thoughts, bizarre IIIA
vision problems, foggy vision
vision focus problems IIIB
cognitive problems, foggy brain II
sexual dysfunction IIIB
muscle weakness II
fatigue, II
low stamina IIIB
apathy, lack of emotions II
balance problems IA
stiffness, slow movement, "walk problems" IIIB
skin, sticky skin, oily skin, O
head, horrible sensations like heavy , feels like I am going to fall, IA
head, pressure in my O
vibrations through body, IIIB
burning sensations (arms and legs) IIIB
alcohol, don't tolerate, and at food like pizza IIIB
trauma of getting through this II
anxiety and depression worse than pre the AD's IIIB
feeling intoxicated IIIB
dizziness in different ways, vertical and horizontal IA
sensitive for sound IB
diarrhoeas, yellowish II
restless legs O
insomnia II
emotionally very numb IIIB
didn´t react to anything IIIA
fatigue, felt more and more tired II
restlessness, severe
anhedonia, disinterested in life
muscles, stiff neck II
stiff spine
anxiety.IIIB
nausea, dizziness, IA
joints, unstable (feet fold when I walk)
muscle weakness. IIIB
nausea, vomiting, dizziness IA
diarrheas per day, 10-12 white to yellowish, full of mucus and sometimes blood IIIB
bleeding and blisters in mouth and nose, blisters on hands and feet O
coughing blood and lots of phlegm and plugs resembling hard plastic O
urinary and fecal incontinence during the first 3-4 months, sometimes difficulties passing urine which is filled with white particles and is dark and reddish IIIB
muscle weakness throughout body, often so severe I couldn´t stand up, IIIB
dropping things I tried to hold, IIIB
joints jumping out of place O
speech, slurred II
muscles, trunk, so weak I could only use neck muscles for breathing IIIB

cold, ice, arms, legs, back, nose and tongue O
blue lips, fingers and toes IIIB
temperature, body, 34 degr. C O
jerking, twitching of muscles visible through skin, involuntary movements IIIB
weight loss, rapid 40 pound IIIB
muscle wasting IIIB
skin and eyes, strange sticky substance resembling resin and not water soluble coming out in, causing intense discomfort and pain, O
skin getting thicker and thicker, is bright red and very painful O
skin itching and burning intensely IIIB
blood pressure often dropping to 85/55 IIIB
skin, layer of water under O
ears, water in, and lungs O
anxiety, panic and fear IIIB
severe confusion IIIB
aphasia IIIB
amnesia IIIB
zaps,"electric" shocks, travelling through body every few minutes IIIB
stiffness throughout body, often unable to move legs IIIB
cramping, twitching IIIB
heaviness in body, feeling of leaden O
heart palpitations and uneven heart rhythm IIIB
tripping, falling IA
feet hanging down, unable to flex toes on left foot
stomach, burping, bloating, gas in, causing lots of pain II
swallowing, food doesn´t move from esophagus down to stomach
body function has just stopped, like every normal. IIIA
vibrations travelling through body IIIB
cramping, attacks of painful IIIB
head is thrown backwards and twists to the left and gets stuck in that position for minutes IIIB
night sweats, intense, NOT relieved but made worse by estrogen supplements I used with good results prior to this IIIB
intolerance to lots of food, especially dairy products, fish, meat, fat, sugar. IIIB
intolerance to drugs and chemicals IIIB
fungal infections in mouth, nose, intestines and lungs. IIIB
severe fatigue IIIB
sensitivity to sound, light, smells IIIB
feeling of gas moving upwards inside body tissues and under skin. IIIB [Formication]
feeling of popcorn popping inside my legs and all over back IIIB [crepusculation]
stiffness persists altnernating with weakness
swaying in all directions, upper body, due to weakness around waist area. IIIB
head upright, difficulties holding my IIIB
stiffness sets in, can walk about one block before compelete IIIB
begin to breathe a bit better
coughing up mucus and sharp, white particles
cough now sometimes dry and painful O
muslces still stiff and making cracking sounds when I move IIIB
breathlessness
muscles, hardening painful and stiff IIIB
intolerance of dairy IIIA
joints a bit more stable but still have to bandage ankles IIIB
sensitive to stress II
sticky skin, this process is physically quite painful. O
insomnia (wired feeling, paired with flushing and nausea) IIIB
nausea IA
anhedonia II
stress intolerance II
muscle pain/stiffness IIIB
joint pain, psuedoarthritis/ IIIB
diminished quality of life. II
weight loss, 25 lbs IIIB

insomnia IIIB
restless legs. IIIB
concentration problems II
memory problems, IIIB
cognitive, inability to find and use the right word in conversations. II
confused and spacey, days/Weeks of feeling .IIIB
anxiety without heart racing IIIB
depression – didn't have before II
sensitive to noises IB
emotional and moody. IIIB
anger / Rage that lasts for weeks.IIIB
headaches for weeks at a time. O
sinuses and allergies worse and more infections. IIIB
asthma acting up. O
eyes are puffy with dark circles. O
menopausal type symptoms that blood work does not support Chills/hot flashes, night sweats. Every month a night of waking up shivering and cold. O
menstrual cycle has changed longer and heaver. O
PMS is a lot worse. IIIB
hypoglycemia O
grinding of teeth. IIIB
fatigue most of the time. IIIA
fear of losing my sanity and not understanding what is going on .IIIA
anhedonia and pain, All I have right now is an existence.
confusion, intense, episodic, daily bouts of IIIB
mood swings (improved but not resolved) IIIB
rage (improved but not resolved) IIIB
tremors, muscle twitches, muscle stiffness, IIIB
joint pain. IIIB
eye twitch (resolved) O
insomnia (improved but not resolved) IIIB
vision problems, episodic IIIB
zap, electric, sensations IIIB
fatigue, profound II
depersonalization
derealization (improved by not resolved) IIIB
motivation problems, profound IIIB
stressors, a cartoonish inability to deal with everyday activities and IIIB
memory problems IIIB
attention problems II
cognitive, information processing deficits, a variety of other II x 5
OCD symptoms (improved but not resolved) IIIB
flu-like symptoms IIIB
nausea, vomiting (resolved) IA
dysautonomia O
light sensitivity
sound sensitivity IB
terror, intense waves of terror/feelings of impending doom (not a panic attack).IIIB
hypersensitivities, severe, to medications, supplements, IIIB
exercise, intolerance to IIIB
hypersensitivities, the sun. IIIB
dry mouth, IIIB
sweating, increased O
night sweats, IIIB
heart palpitations, IIIB
abdominal pain and cramping, II
belching, bloating, II
diarrhea, dyspepsia, flatulence,
heartburn, indigestion II
nausea, IA

toothaches, O
asthenia, weakness II
chest pains, IIIB
flatulence, heartburn, Indigestion IIIB
chills, IIIB
fever, IIIB
hot flashes IIIB,
leg and extremity pains, IIIB
malaise, II
flu-like symptoms, IIIB
tightness in my chest, IIIB
unexplained bruising, O
arthralgia (sudden sharp nerve pains in the joints), IIIB
horrible upper and lower back pains, IIIB
jaw pain and stiffness, IIIB
serious thirst, IIIB
dizziness and IA
headache, migraines, O
tics, IIIB
tremors, IIIB
twitching, IIIB
vertigo, IA
rage and agitation, IIIB
anxiety attacks, IIIB
apathy, IIIB
impaired concentration ability, II
impaired memory, IIIB
depersonalization, IIIB
depression, IIIB
terrible dreams, IIIB
abnormal crying, IIIB
insomnia, IIIB
suicidal thoughts ,IIIA
acne, eczema, dry lips and skin, IIIB
boils and hives, IIIB
fear, IIIB
and rashes. IIIB
Still doing well, no smoking yet and I haven't been hospitalized
foggy brain II
hypothyroidism O
costochondritis O [rib cartilage]
paresthesia, left arm and back [numbness] IIIB
sound sensitivity IB
chemical sensitivity IIIB
attempted suicide
severe head and neck pain. IIIB
severe depression and IIIB
anxiety unlike anything I had IIIB experienced before taking SSRIs,
chronic insomnia, IIIB
akathisia, IIIB
dysautonomia including IIIB
rapid, uneven, or forceful heart beat, IIIB
detox rashes, II
muscle pain II
joint pain and IIIB
injuries from muscle spasms ,II
impaired circulation, IIIB
disrupted menstrual cycle, IIIB
hair loss, IIIB
gastrointestinal problems, II

increased "floaters" in eyes, IIIB
hypersensitivity to supplements IIIB and
medications, IIIB
cognitive impairments, IIIB, IIIB
quasi-OCD, etc. IIIB
immune system. lowering resistance of the, IIIB
strange and unwelcome thoughts IIIB
visual and olfactory hallucinations IIIB
depression II
anxiety for no reason IIIB
sometimes I forget to breath O
fatigue II
zaps IIIB
wooshes IIIB (went away and reappeared from month 9 to present)
lethargy II
rages so violent I broke mirrors or threw whatever was in my hand to punching holes in my walls. IIIB
panic attacks where I could not breath IIIB
dizziness, extreme. I had to hold walls to walk sometimes. Had to give up hiking. Elevation made it 10 times worse.[oxy deprivation] IA
no memory. II
head zaps. IIIB
stressed, easily. IIIB
slower mentally. IIIB
concentration, no, so I could not watch tv or read things of length II
clumsiness, had bruises all over my body from walking into things IA
fell of my treadmill a few times IA
tinnitis, which lasted well after my w/d stopped IB
insomnia. IB
jerks, arm, leg and sometime whole body jerks IIIB
shakes, body IIIB
obssesive thinking IIIB
muscles, aching IIIB
energy, no II
inability to work; I could not drive, cook or shop. Husband did everything for a while.
taste and smell was gone. II
food intolerance; I could not eat red meat. O
sound sensitivity IB
nausea II
diarrhoea. II
mood swings, severe. IIIB
anxiety and depressions that kept me in bed. II
post traumatic stress syndrome [PTSS]. IIIB
suicidal thoughts. IIIA
psychosis that let self harm. IIIB This had nothing to do with the suicidal thoughts.
anxiety II
panic IIIB
depression, IIIB
manic episodes IIIB
agoraphobia II
suicidal ideation IIIA
many other physical symptoms. IIIB x 6
cognitive/memory problems which cost me a news editing job II
speech problems which turned my career away from doing live radio reports II
depersonalization to the point I don't recognize myself IIIB
zoning out - missing exits in traffic, II
disorientation; forgetting where I was II
anxiety pain in chest IIIB
heartbeat, rapid IIIB
psychosis; lucid nightmares - can't tell if I'm awake or sleeping IIIB
trembling IIIB

twitching nose and face II
muscle soreness and knots II
skin rash IIIB
head, sharp, stabbing pains in II
zaps IIIB
blurred vision IIIB
dizziness IA
nausea IA
ongoing diarrhea IIIB
exhaustion II
irritability II
restlessness and skin-crawling sensations IIIB
relentless insomnia IIIB.
anxiety IIIB
balance problems IA
depression II
tinnitus IB
apathy II
loss of confidence II
scary thoughts IIIB
no patience IIIB
no humour IIIB
depersonalisation IIIB
no appetite IIIB
feel like I am just existing from one day to the next II
memory loss IIIB
depersonalizaiton, severe IIIB
derealization, IIIB
insomnia, so severe that it was not uncommon to go 3-4 days without a wink of sleep. (Has improved about 80%) IIIB
depression II
suicidal urges IIIA
hopelessness and severely impaired confidence. and outlook for the future IIIB
emotional disconnection and difficulty relating to others IIIB
speech prosody; For instance, it is still hard to talk to friends on the phone. My speech is stilted, my emotions stifled and I lack spontaneity. IIIB
speech prosody; consequently, I sound "robotic" II
numbness and tingling, frequent, throughout my body, but particularly in my right arm and hand. IIIB
fear; "Pharmaceutiphobia" -- A distinct, deeply-rooted fear and mistrust of doctors and the entire medical community, particularly pharmaceutical companies and everyone involved in the research, development, marketing, and dispensing of drugs.
memory is still dreadful and IIIB I have been off Seroxat for 18 months.
Intrusive thoughts; I did have those awful thoughts IIIB
balance problems IA
memory loss IIIB
concentration problems II
energy, no II
dreams, vivid IIIB
insomnia II
anxiety much worse than before Paxil, esp in the morning IIIB
depression much worse than before Paxil IIIB
PMS, terrible IIIB
headaches, usually very sudden O
dizziness IA
zaps IIIB
stomach, upset, diarrhea, II
nausea IA
heartburn II
stomach sensitive to many foods IIIB
sensitive to medications IIIB
anger, outbursts of IIIB
agoraphobia II

crying spells IIIB
muscles, achey II
heart palpitations IIIB
breathing, trouble, at times IIIB
acne IIIB
hair falling out IIIB
feeling faint IIIB
sensitive to bright lights and IIIB
sensitive to loud noise IB
vision blurry and trouble focusing eyes IIIB
circulation problems, feels like my blood is too thick IIIB
ears, pain in my IB
ears, ringing in my IB
muscle twitching II
weight gain IIIB
personality changes IIIB
motivation, no IIIB
memory loss IIIB
stomach problems IIIB,
dizziness, headaches IA
sleeping, excessive IIIB
eating, excessive IIIB
dreams, vivid IIIB
zaps
anxiety, nightmarish
suicidal impulses IIIA
violent impulses IIIB
mood swings, severe IIIB
sex drive, loss of IIIB
derealization IIIB
depersonalization IIIB
hallucinations IIIB
crying spells IIIB
depression, severe IIIB
memory, short-term, loss IIIB
concentration, poor II
sensitivity to noise IB
insanity; constantly thought I was going insane IIIB
OCD far worse than before I went on Celexa IIIB
panic, severe day-long IIIB
psychosis; visions, strange dream like, that occured at random while awake IIIB
eye; constant stabbing pain behind my right eye O
migraines O
joint pain IIIB
muscle pain and stiffness , severe II
heart palpitations IIIB
hair loss IIIB
zaps; electric shock sensations throughout my body IIIB
eyes; "snowy" vision IIIB
eyes; spiderweb floaters in vision IIIB
vision, after-images in IIIB
vision, smeared, when turning head IIIB
twitching IIIB
body jerks, involuntary movements- IIIB
dizziness IA
tinnitus IB
fatigue II
muscle weakness II
skin in my legs, sensation of bubbles popping under my O
skin, crawling sensations IIIB

parathesia IIIB
skin; thick substance (callus-like) developed on my palms O
bloating; severe IIIB
jaw-clenching IIIB
thyroid, a warm nodule in my thyroid
endometriosis
ovarian cyst, ruptured
muscle twitching II all day
pain, occasional IIIB
cognitive; inability to think clearly or process Information II
memory; forgetfulness (short term) II
nightmares, vivid - (mostly recurring of car crash) IIIB
twitching - legs and eyes II
rage, extreme - "short fuse" IIIB
derealization; feeling of unreality IIIB
emotional disconnection from other human beings including my children - like I don't want to talk to ANYONE IIIB
fatigue; extreme tiredness II
anxiety / fear of something terrible about to happen IIIB
dizziness (like low blood pressure) IA
buzzing / Itching in one ear IB
sensitivity to loud noise IB
sensitivity to bright lights IIIB
sensory "overload" IIIB
depression, intense II
zaps IIIB
rage, extreme bouts of IIIB
disconnection to the feelings of others and life IIIB
sensitivity to light/IIIB
sensitivity to noise/ IB
sensitivity to fluorescent lights IIIB
memory; forgetfulness II
insomnia IIIB
joints, pain in, especially hands IIIB
cognitive; could not spell or type words out Properly II
crying spells that lasted for hours IIIB
mood swings, severe IIIB
suicide, thoughts of IIIA
anxiety and fear IIIB
concentration, terrible II
apathy II
humour, loss of IIIB
hair loss IIIB
muscle pain II
jaw clenching II
energy, loss of II
nausea, severe IA
vomiting, heaving, gagging on a daily basis. IIIB
appetite, no IIIB
weight loss, severe, (37 lbs) IIIB
digestive problems: distended stomach, bloating, gas, eructions, IIIB
diarrhea, (explosive) IIIB
swallowing, food not going down, difficulty IIIB
sweats, hot, day and night, IIIB
cold chills
cold extremities IIIB
fatigue, severe IIIB
skin; greasy face O
dizziness all day IA
shortness of breath v IIIB
thirsty, extremely IIIB

nerves; crawling sensations inside head and going down to face. IIIB
muscle weakness, walk with cane. II
eyes; floaters in my right eye IIIB
work; can't cook, drive, shop IIIB
jaw clenching II
body tremors, especially my hands IIIB
anxiety II
cognitive; brain fog II
vision, blurred IIIB
memory loss, short term, memory lapses II
cognitive; spaced out II
concentrate, can't focus, slowed thinking II
cognitive; can't function "normally" II
sound sensitivity, extreme, sounds seems to be magnified 1000 times. [hyperacusis] IB
insomnia IIIB
disconnection from the outside world IIIB
depression due to all of the above. IIIB
hallucination, intrusive audio; In hospital Effexor XR was tried and since then got music (tunes) in my head. Effexor was withdrawn there, but tunes keep on playing. Sleep is the only relief. IIIB
sensitive to smells, extremely, food smells awful, as do all chemicals and smells that I never noticed prior to withdrawal IIIB
inability to chew food II
taste perversion II
suicidal thoughts and ideation [hospitalized for] IIIA
nausea, severe, upon awakening that won't go away until I fall asleep IA
vomiting with some food, dry heaving, heaving with some water, gagging daily.
burping all day, no relief, feel as if I have to vomit all day long.
flatulence, no relief II
diarrhea and loose stools. Benefibre doesn't help. IIIB
appetite, no
anorexia; aversion to food and drink, struggle to eat and drink, have to force myself. Can barely eat, resorting to purred & liquid foods, very little variety I can have. Drinking Ensure since Jan 2006. IIIB
swallowing; feels as if food is not going down, bending over makes me more sick, hurts II have to wear loose clothing otherwise it hurts from the throat down to the end [vagus nerve tract].
esophagus; pain from throat down to the end. II
jaw; trouble chewing
esophagus; trouble swallowing. II
feel more sick after eating.
urination, frequent; pee too often. IIIB
sweats, hot IIIB
sweats, ice cold IIIB
breathing; lack of air, especially in the morning. IIIB
cannot lie down after eating for hours on end, have to sit up all day even though I'm exhausted mentally and physically.II
fatigue; exhausted mentally II
fatigue; exhausted physically II
heart beats faster. IIIB
trembling of body, especially legs and hands II
sound sensitivity, severe IB
hallucination, auditory; "music" drones in my head and amplifies (magnifies) through everything that has a sound
hyperacusis; amplifies (magnifies) through everything that has a sound; shower, fridge, water taps, vents, heating and air conditioning, car engine etc – EVERYTHING!! IIIB
olfactory disorder. Food, chemicals, detergents, soaps, flowers, grass - all smells are horridly magnified. (unbearable)
Yes vinegar too, wine etc. unscented products still smell awful. II
thirst, excessive, yet can only sip water IIIB
metallic taste in mouth. II
taste perversion. II
hypersensitivity to sensory stimuli, autistic degree of; no rest or peace due to the above, II
agoraphobia; housebound, cannot go out due to the above
anxiety due to the above. II
life, no quality of, whatsoever. IIIB
dysautonomia IIIB

sensitivities / intolerances, drug IIIB
saliva, excess O
anger, extreme , irritability IIIB
anxiety & depression (much worse than anything I experienced before taking SSRIs) IIIB
muscle weakness/stiffness II
fatigue II
appetite, huge I get hungry very quickly. Related to blood sugar perhaps? IIIB
skin, very dry skin on face & scalp. I had oily skin before discontinuing SSRIs. IIIB
restlessness IIIB
memory problems II
cognitive problems II
stress, low threshold IIIB
dizziness, occasional IA
co-ordination problems II
digestive problems, constipation II
nerves; pins and needles in arms and legs IIIB
feelings; how bad the experience has been. IIIB
anger
weight, gained 40 pounds IIIB
zaps IIIB
flu-like feeling IIIB
dysautonomia IIIB
panic attacks, major IIIB
tachycardia IIIB
blood pressure, high IIIB
breathing, trouble IIIB
IIIB
heartburn, severe IIIB
abdomen, pains in my. IIIB
IBS IIIB
tachycardia, severe IIIB
restlessness IIIB
fatigue II
sweats, hot and cold IIIB
feelings; It felt like I was going to die.
chest and back pain, severe [hospitalized] IIIB
pain; severe back, neck, shoulder and chest pains daily IIIB with absolutely no let up.
sleep, poor. IIIB
stress increases pain IIIB
suicidal depression IIIA
tachycardia IIIB
weight gain IIIB
euphoria IIIB
weight gain IIIB
anhedonia; blunted emotions IIIB
depression, increased IIIB
panic attacks, severe IIIB
tachycardia IIIB
breathing, trouble IIIB
zaps, brain IIIB
flu-like feeling IIIB
sweats, hot and cold IIIB
blood pressure, high IIIB
tinnitus; ringing in my ears IB
head, pressure in my O
muscle pain , severe IIIB
neurological pain - electric tingling on left side of body IIIB
back pain, lower, severe IIIB
dizziness IA
vertigo IA

depression IIIB
suicidal ideation IIIA
anxiety IIIB
muscle twitching II
headaches O
heart beats, ectopic (skipped) IIIB
mitral valve prolapse (confirmed) IIIB
disabled, unable to work.
psychological disorder diagnosis, 17 new IIIB x 17
suicide attempts, 5 documented hospitalized IIIA x 5 (never even once thought of the idea of suicide prior to withdrawal)
heart issues IIIB
dizzyness [to ER] IA
headaches O
vomiting, daily IIIB
weight loss; lost 50lbs below healthy normal IIIB
hallucinations, visual IIIB
delusional, IIIB
paranoid, and had psychotic episodes IIIB
self harm (cutting- never before even thought of that) IIIB
tachycardia IIIB
zaps, brain O
flu-like feeling at all times IIIB
sweating, extreme night IIB
tinnitus; ringing in my ears IB
pressure in my head O
back pain, lower, severe IIIB
dizziness IA
vertigo IA
anxiety IIIB
muscle twitching II
headaches O
ectopic (skipped) heart beats IIIB
mitral valve prolapse IIIB
whoosh IIIB and drag feeling when i look from side to side is always present now
symptoms, many more IIIB
amnesia IIIB
self-harm] IIIB
anxiety and IIIB
panic attacks, which i had never had except since w/d, IIIB
headaches, O
anxiety to a lesser degree, IIIB
irrational thoughts, IIIB
fear of death, IIIB
depression; everything being wrong / and also IIIB
memory problems II
gastro-intestinal discomfort/bloating II
restless legs, arms sydrome IIIB
rage IIIB
violent rage IIIB
suicidal thoughts/planning IIIA (even though I have no desire to die)
disassociation IIIB
weepiness IIIB
moods, exaggerated IIIB
ear; left ear feels closed at times-IA
tinnitis IB
stress, unable to deal with IIIB
memory, long term, is really foggy, IIIB
short-term memory
emotions; stifled, overcompensating- I don't know. IIIB
hallucinations, audio

dreams, vivid; excessive dreaming/images & talking in my head when I close my eyes. IIIB
hallucinations, visual
self-control, loss of; I cried all the friggin' time about it IIIB
insanity, fear of ; because it literally felt like I was losing my mind. It's slowly resolving. IIIB
insomnia IIIB
OCD type thoughts that sometimes get better or worse. IIIB
anger and irritability for no reason IIIB
crying for no reason. IIIB
restless/aching legs IIIB
anxiety, morning-afternoon IIIB
emotions; inability to actually feel any emotion whatsoever for days. IIIB
confusion II
speech; sometimes having trouble talking II
cognitive processing; have trouble understanding my senses IIIB
speech; ;sometimes mix up my words (like I'll say basement instead of trunk of the car?) IIIB
depersonalisation IIIB
cognitive processing; feel distanced from my thoughts, or that my memories feel like someone else's IIIB
depersonalization; sometimes when someone calls my name it feels weird to me.. IIIB
very loose stools II
vision; moving static in my vision all the time and a lot of floaters IIIB
headaches, intense, that last like a minute. Not a regular symptom. O
circulation is worse than it was before. IIIB
unable to warm up after coming inside for literally hours. IIIB
pressure feeling on my forehead. It's more like on the skin. I get it a lot more often at night. O
twitching in one eye II
social problems IIIB
mood swings IIIB
depression, episodes of extreme IIIB
suicidal thoughts IIIA
fatigue; tiredness IIIB
tremor in legs IIIB
muscle and joint pain IIIB
irrational and paranoid thinking IIIB
over-reacting and highly sensitive IIIB
memory loss, short term II
tingling /numb feet and hands IIIB
appetite loss, total, and then sudden bursts of never feeling full. IIIB
dyslexia, pseudo, changes in speech, II
speech prosody; tone [of voice] gets messed up II
anger IIIB
body temperature, dys-regulation IIIB
depersonalisation IIIB
exercise, unable to IIIB
no longer play guitar, perform or write music IIIB
listening to music is sometimes a Problem IB.
socialize, unable to II
hobbies or interests, unable to pursue II
work, unable to IIIB
house bound most of the time IIIB
quality of life... Extremely Poor. IIIB
pressure in head..O
headaches O
foggy Brain II
heartbeat pounding in my head O
hallucination, audio; h.f.[high-frequncy?] white noise in head IB
eyes and sockets, painful IIIB
pains in face II
mouth; tingling tongue and roof of mouth feels raw. IIIB
taste, metallic IIIB
jaw tension IIIB

mouth; strange teeth pressure sensations, IIIB
mouth; teeth sensitivity.. unrelated to dental problems. IIIB
derealization, severe IIIB
depersonalization. IIIB
breathing; feelings of suffocating or drowning. IIIB
inability to relax or feel calm IIIB
panic attacks IIIB
anxiety IIIB
depression, intermittent bouts of IIIB
suicidal thoughts. IIIA
sexual dysfunction. IIIB
fatigue; very little mental or physical stamina II
muscle weakness. II
circulation problems. IIIB
energy, very little to no
fatigue, bouts of extreme IIIB
concentration, poor II
memory problems IIIB
anhedonia, unable to feel pleasure, joy, happiness etc IIIB
creativity, no IIIB
body tremors, vibrations, shaking IIIB
muscle spasms, muscle rigidity II
pains, shooting
electrical feelings, buzzing and, throughout my body IIIB
restlessness, inner IIIB
agitation extreme
akathesia IIIB
restless leg syndrome IIIB
intolerance to medication, IIIB
intolerance to alcohol, coffee etc. IIIB
intolerance to various sensitivities IIIB
stress, intolerance to IIIB
insomnia, sleep disturbances IIIB,
nightmares. IIIB
sleeplessness, only average 4 hrs sleep a night.
vision, very poor IIIB
vision, blurred IIIB
agoraphobic, but no fear of open spaces or being outside.. IIIB
rash on neck
chest pains IIIB
numbness in chest, IIIB
numbness in hands, etc
stomach pains and sensations.II
acid sensations in chest and stomach after eating.
numerous body pains. IIIB
anal bleeding O
body temp fluctuations. IIIB
feelings of anger, bitterness, resentment. IIIB
head pressure, O
vision problems, IIIB
feeling intoxicated, IIIB
numbness and tingling especially arms, IIIB
parasthesia, IIIB
apathy, IIIB
sex; loss of libido IIIB,
stress, IIIB,
sensitivty to chemicals IIIB
intolerance to stimulants, drugs IIIB
sound , intolerance to IB
light, intolerance to IIIB

gastric problems, II
sleep disturbances IIIB
agrophobia (hard to go out on my own due to horrible scary symptoms), IIIB
anxiety, IIIB
panic, IIIB
some depression due to feeling so crap, IIIB
breathing; body feels starved of oxygen, IIIB
breathing problems, IIIB
chest pain, IIIB
agitation IIIB
mood swings IIIB,
anti social at times, IIIB
feel unwell IIIB and unacceptable to how i was pre seroxat
very suicidal IIIA
dizzy IA
nauseas IA
anxieties, large increase in IIIB
depression - mild to severe IIIB
anxiety - mild to severe IIIB
stress, extreme sensitivity to IIIB
sensitivity, extreme, to any perceived loss IIIB
irritability IIIB
emotional outbursts IIIB
social uneasiness IIIB
agoraphobia IIIB
vision, blurred vision IIIB
sinus issues IIIB
muscle pain and spasm II
fatigue even though I exercise 6 days a week IIIB
weight gain IIIB
urinary issues - clean urine and normal PSA O
memory problems IIIB
cognition; unable to think through complex Thoughts IIIB
shaking and sweating IIIB
weight gain, and I was underweight IIIB
memory problems, horrible IIIB
cognition; poor word retrieval, IIIB
cognition; difficulty understanding and having conversations now, IIIB
sex drive, lowered IIIB
cognition, slowness, difficulty understanding conversations and IIIB
memory loss, short-term IIIB i literally do not remember what I said the minute before.
dizziness, severe - 8 weeks and was daily. Wrecked my vehicle twice because of the dizziness. IA
intrusive and scary thoughts IIIB
hallucinations, visual IIIB
tinnitus, it's extremely loud IB IB
insomnia IIIB
rage IIIB
anger issues IIIB
anxiety IIIB
buzzing deep inside my body IIIB
adrenaline surges on a regular basis IIIB/
disequilibrium; can't even walk straight IA
work; let alone hold a job. II IIIB
vertigo/dizziness IA
intolerance for stress, physical IIIB
intolerance for stress, emotional IIIB
anhedonia IIIB
depression IIIB
dark holes ,regular IIIB
crying spells, can last for hours on end IIIB

suicidal ideation IIIB
intrusive thought IIIA
tachychardia IIIA
heartbeats, missed IIIB
fear of medicine and distrust of doctors IIIB
fatigue, bone-crushing fatigue IIIB/II
immune system instability – Since the cold turkey withdrawal, I have had colds, IIIB/IIB
sinus infections, IIIB
bladder infections, IIIB
stomach viruses, and IIIB
rashes IIIB
immune system IIIB
poor memory IIIB
cognition problems IIIB - this was severe IIIB
cognition issues II
pain, chronic, facial IIIB
pain, chronic, left eye IIIB
depersonalization IIIB
derealization IIIB
fatigue, not stress-related II I often feel it when I am well rested
anxiety II
agoraphobia II
anxiety IIIB
pain, severe, constipation II
concentration, severe lack of ability II
feeling like a zombie; II
indifferent to situations in life II
personality changes II
forgetfulness II
lethargy II
dizziness IA
sleepiness all the time) IIIB
less active IIIB
anxious II
depressed II
sensitivty to noise IB
emitional outbursts IIIB
inability to think comples thoughts IIIB
concentrate, inability to II
eyes, excessive irritation in IIIB
mood swings IIIB
vision, odd IIIBmuscle twitches involuntary II
myoclonus [muscle spasms] II
dizzy spells IA
eye; ocular migraines IB
muscle pain II
memory problems IIIB
panic attacks IIIB
fear & anxiety, intense, with no trigger and unrelated to environmental factors IIIB
severe physical agitation IIIB
mental agitation IIIB
suicidal thoughts IIIA
depression, severe bouts of IIIB
insomnia IIIB
intrusive thoughts IIIB
obsessive thinking IIIB
leg aches & pains II
inner restlessness II
crying for no reason IIIB
heart beating over-rapidly IIIB

anxiety induced by exercise IIIB
strong suicidal feelings out of the blue & for absolutely no reason IIIA
jerking II
sensitivity to light IIIB
crowds, sensitivity to IB
elation when listening to music IB
feeling high IIIB
paranoia IIIB
electric zaps IIIB
self-harm, desire to, was very strong in early withdrawal IIIB
emotions; extreme reactions IIIB
intolerance to stress IIIB
immune system, impaired IIIB
aggression and rages IIIB
depression, extreme severe IIIB
sleep disruption IIIB
obsessive thoughts, severe bouts of IIIB
extreme restlessness, IIIB
feeling like I had been obliterated IIIB
anxiety II
intrusive thoughts IIIB
obsessive thoughts IIIB
depression IIIB
insomnia & sleep disturbance IIIB
leg pains II
paranoia IIIB
headaches – head sometimes feeling heavy when exercising O
emotional sensitivity IIIB
stress intolerance IIIB
suicidal thoughts IIIA
derealisation & IIIB
depersonalisation
disconnected to other people, physically and mentally. IIIB
immune system, impaired IIIB
diarrhoea II
aggression & rages IIIB
memory difficulties II
concentration difficulties II
memory, bouts of bad IIIB
cognition; difficulty making decisions IIIB
labile moods IIIB
vertigo IA
dizziness IA
fatigue II
mania, 20 cups of espresso {MANIA] IIIB
psychosis, and 4 hits of LSD.[PSYCHOSIS]. IIIB very scary.
hallucinations, intense IIIB
panic attack, second major, awoken from my sleep. IIIB. lasted for about 45 minutes,
numbness and weakness in arms IIIB and hands, legs and feet.
intense fear and anxiety IIIB
sinus infection and bronchitis IIIB
panic attacks again IIIB
diarrhea II
nausea IA
parasthesis of the arms and legs IIIB
vision, weird IIIB
dizziness symptoms IA
waking up thinking I am going to go completely insane IIIB
tinnitus, though not as severe[as] when on Paxil. IB
severe, unnatural anxiety IIIB

agitation IIIB
vibrating feeling II
repetitive "squirts" of anxiety deep in abdomen IIIB
severe insomnia- did not sleep for days in the first 5 months IIIB
depersonalization, severe, verging on psychosis IIIB
suicidal ideation, husband would spend days monitoring me IIIA
wooshing feeling when turned head IA
disequilibrium when standing, walking IA
time, altered sense of time IIIB
skin, itching- scratching produced red, raised welts IIIB
hopelessness IIIB
depression, severe IIIB
menstrual cycle disruption IIIB
derealization, feelings of unreality IIIB
feeling, "vacant" feeling- I describe it as having my "hard drive" erased IIIB
fear, unprovoked, irrational, overwhelming sensations of fear IIIB
paranoia, severe IIIB
intrusive thoughts, very scary IIIB
dreams, vivid, unsettling dreams; nightmares when I was able to sleep IIIB
feces lost pigment for awhile IIIB
tingling sensation on face- specially around lips IIIB
hypertension IIIB
tachycardia IIIB
concentrate, unable to II
memory, poor II
headache O
anhedonia IIIB
hallucinations, auditory, twice - heard loud bang like gunshot IB
anorexia- lost 20 pounds from thin frame IIIB
crying spells that I could not control IIIB
sensitive to light, IIIB
sensitive to sound IB
zaps, mild electrical - I had very few IIIB
intolerance for stress IIIB
numbness and tingling in feet- difficult to flex toes. Left foot worse than right. IIIB
anxiety - perceived primarily in abdomen II
insomnia - improving- now sleep most nights but only for a couple of hours of varying quality IIIB
depersonalization, severe - minimal improvement IIIB
itching IIIB
hopelessness IIIB
feelings of unreality-IIIB
fear- some improvement IIIB
time, altered sense of, improving IIIB
hypertension- improving IIIB
tachycardia- improving IIIB
concentrate, unable to II
memory- improving II
anhedonia IIIB
crying spells improving IIIB
sensitive to light, improving IIIB
sensitivity to sound- improving IIIB
pressure in head and left temple O
numbness and tingling in feet- difficult to flex toes. Left foot worse than right. little to no improvement IIIB
arthritis in knees- have no history of arthritis before IIIB
vertigo, VERTIGO! IA vertigo, dizzy spells, pretty severely where I have to pull over if I am driving- these bouts can happen every day, several times a day or once every few days no rhyme or reason to them.
constipation / diarrhoea II
tingling in fingers and toes IIIB
twitches in leg and feet IIIB
head numbness O

concentrating, difficulties II
dehydration O
headaches O
pains, sharp, in upper right thigh and hips IIIB
twitches / tics in head IIIB
numbness in head
sensitivity to lights and IIIB
sensitivity to sounds IB
concentrating, difficulties II
speaking, difficulties II
crying for no apparent reason IIIB
intolerance to certain vitamins IIIB
intolerance to cafeine, alcohol IIIB
intolerance to medications IIIB
intolerance to foods IIIB
heart palpitations IIIB
blushing for no apparent reason IIIB
dreams, weird IIIB
sleeping, difficulties IIIB
exercise, stressed and overwhelmed when doing high impact exercises IIIB
difficulties sleeping at night
low mood/lack of motivation IIIB
head tics expecially if I didn't sleep well the night before
numbness in head
numbness around mouth IIIB
tingles or sharp pain in fingers and toes IIIB
pain in the upper thigh/hip area around the time of my period IIIB
sensitivty to crowded/loud places. IB
tingling in hands and feet. IIIB
zap, brain IIIB
nausea. IA
insomnia. IIIB
concentrating, difficulty II
depression II
chest pain IIIB
anxiety worse than anything I'd ever experienced, including panic [in hospital for a week] IIIB
tingling in hands and feet. IIIB
nausea and vomiting. IA
insomnia. IA
difficulty concentrating. II
some depression. II
horrible anxiety again, including episodes of panic. \ IIIB
muscle twitches. IIIB
tremor. IIIB
feeling like one big, raw, crackling nerve. IIIB
anxiety / panic IIIB
emotional confusion, IIIB
sick stomach IIIB
nausea IA more so than during tapering or withdrawals, the actual "withdrawal" was more physical than emotional but now its emotional.
panic attack, I went from having one panic attack a day to having 10 IIIB
vertigo IA
dizziness/ IA
work; I am incapacitated 24/7
feelings; not one minute of the day when I feel normal.
brain felt like it was plugged in to an electrical outlet
disequilibrium IA
GERD II
stiff muscles (legs) II
extreme anxiety IIIB

depression IIIB
strange head feelings O
seizures IIIB
panic attacks IIIB
phobia, all kinds IIIB of
throat/neck issues IIIB
vision problems IIIB
chills IIIB
thoughts, weird, frightening IIIB
visual Disturbances IIIB
palpitations IIIB
dizziness IA
anxiety, elevated IIIB
tinnitus IB
depersonalization IIIB
tingling, buzzing, and sometimes weakness II
ear; left ear has also felt blocked.IB
left-sided feelings and my ear at that point. IIIB
gradually increasing in anxiety, totalIIIB
insomnia, and IIIB
profound confusion IIIB
accompanied by inability to eat, IIIB
extreme exhaustion. II
partially paralyzed, IIIB
vomiting, IIIB
intense suicidal thoughts IIIA
jerking movements, IIIB
akathisia, II [ACATHISIA]
blackouts, IIIB
seizure activity, IIIB
memory loss, IIIB
skin boils, O
extremely dry skin, O
extreme weakness, IIIB
vision problems with right eye (non constricting pupil) II
right side of face 'falling', II
lips and mouth lost tone and hang II
right foot dragging II
right arm, hand curling II
very waxy/pale skin II
strange waxy hair tone O
hormonal changes (much shortened periods, shortened to 17-21 days) II
depressive episodes IIIB
extreme sensitivity to light IIIB
extreme sensitivity to sound IB
extreme sensitivity to any kind of stress IIIB
great difficulty in speaking to other II
persons due to intense confusion, II
heart palpitations / heart drop, IIIB
intense and varied headaches O
strange mood lability. IIIB
partial complex seizures. IIIB
movement problems as due to Drug reaction IIIB
dystonia II
dyskinesia IIIB
parkinsonism
seizures related to such. IIIB
great difficulty concentrating, II
depersonalization /IIIB
derealization IIIB

dystonia/ II
dyskinesia, II
hormonal trouble, II
asthenia, on and off IIIB
mood lability IIIB
memory problems II
fragmented thoughts II
depression, IIIB
suicidal ideation IIIA
mood swings, severe IIIB
memory is very bad... IIIB
brain fog IIIB
concentrate, inability to IIIB
conversation, cannot have one with someone, IIIB
feeling out of it, IIIB
dizziness, IA
fatigue, II
insomnia IIIB (but not as severe as going through the withdraw)
extreme dizziness IA
loss of balance IA
vertigo (I would be standing still and it felt like my body was moving) IA
anxiety, EXTREME IIIB
depression IIIB
feeling like I was going crazy all day IIIB
obsessive thinking IIIB
agoraphobia IIIB
feeling all day of impending doom IIIB
mood swings, extreme
rage / anger, IIIB
crying IIIB
hallucinations IIIB
skin, burning, tight skin IIIB
skin was crawling IIIB
vision problems IIIB
nausea (Nothing helped this nausea...it was very intense!) IA
head, pressure in my, (felt like my head would explode) O
diarrhoea, 18 months of severe intestinal IIIB symptoms I woke up every morning with severe abdominal cramps and would have that throughout the day...could not be away from the bathroom or go anywhere
breathing, feeling like I could not breathe IIIB
zaps, brain, (felt like a bolt of lightning would hit me in the head out of nowhere) O
akasthasia IIIB
numbness and tingling in extremities IIIB
hair loss IIIB
weight loss IIIB
shaking IIIB
heart palpitations and racing heart IIIB
insomnia IIIB
restlessness IIIB
muscle pain II
muscle twitching II
feeling like my body was vibrating inside IIIB
nightmares, horrible IIIB
suicidal thoughts IIIA
zoned out a lot IIIB
memory problems which continue IIIB to this day
balance problem IA
depersonalization IIIB
sensitivity to light IIIB
sensitivity to sound IB
hypoglycemia O

chest pain and tightness IIIB
sweating, unusual IIIB
mental breakdown. IIIB
homicidal, fight those thoughts. IIIB
anhedonia IIIB
SSRI withdrawal.
insomnia IIIB
weight gain or weight resistance IIIB
extreme anxiety, panic, and fear IIIB
mood swings IIIB
irregular menstrual cycles II
hair loss IIIB
rash on my face, and arms IIIB
digestive system interrupted II
extreme craving of sugar and
carbohydrates IIIB
body zaps and shocks when I try to fall asleep, resulting in O
insomnia IIIB
feeling like I'm on an elevator when I try to fall asleep IA
body zaps and shocks when I wake up IIIB
palpitations when standing up or sitting down IIIB
nerve pain and II
dilated pupils when trying an SSRI again O
shortness of breath, having to think about breathing to not go crazy II
depersonalization IIIB
sensitive to vitamins and herbs and other supplements O
weird, frightening thoughts IIIB
hormonal changes II
feeling of weakness II
racing heart IIIB
chills in legs IIIB
sleep; losing the ability to sleep easy and feeling too awake IIIB
burning sensations that feel deep inside me to my symptoms above O.
sleep disturbances: I've lost the ability to sleep deeply
racing thoughts; brain seems to be racing
panic IIIB
dreams, vivid, often violent dreams In the morning, I feel exhausted and drained.
agitation IIIB
joint stiffness and pain IIIB
eyes; brain zaps and shivers when I move my head or eyes, IIIB
muscle spasms, and brain zaps. II
sound; zaps, brain, triggered by sounds IB
zaps, brain, triggered by sounds IB
anhedonia --this is the worst. I have lost my ability to feel pleasure. IIIB
sexuality;I have absolutely no interest in sex
listening; to music, My life-long love of music is gone
depression, extreme severe IIIB
anxiety, severe IIIB
work; unable to work or care for herself,
agoraphobia, II
severe weakness IIIB
unable to eat IIIB
intolerance to vitamins and
suicide, attempted to end her life at this facility because she was suffering so badly IIIA
stroke IIIB
gut; killed off all my gut flora.IIIB
skin; psoriasis IIIB
irritability IIIB
mood swings IIIB
anger IIIB

manic behavior IIIB
derealization; increasingly no longer cared or took care of things that were happening around me IIIB
fatigued II
work; could barely work and income plunged to below poverty
mania; 9 months and 45 credits with a 4.0 average I decided I had to re-think how fast I was doing this. IIIB
cry, lost my ability to cry, IIIB
feelings; lost ability to care about anything IIIB
emotionally, destroyed IIIB
body, generally, physically destroyed IIIB
anxiety, extreme IIIB
adrenal rushes, constant IIIB
suicidal ideation IIIA
depression, deep IIIB
crying a lot IIIB
anhedonia; not able to feel any pleasure IIIB
concentration, seriously impaired II
intolerance of stress; inability to handle bad news at all IIIB
shakiness II
clumsiness II
depression; the depression is constant and deep, but not suicidal. IIIB
atigue; can sleep, but now always tired II
brusing, unexplained, on arms and legs..i had this on paxil as well..never before. IIIB
vision, blurred vision in right eye II..
heart; atrial flutters IIIB
chest pain. II
back pain. II
pressure in my head between my eyes O
pressure in my cheekbones. II.
anxiety IIIB...
mucous in throat.II.
diarrhoea II
zaps, brain, intense brain zaps at regular IIIB intervals
nausea, extreme II
hallucinations, visual, black shadows, cats, people, constantly; always had to look again IIIB
facial tics / twitches; arm tics / twitches II
shaking, noticeable IIIB
eyes; floaters IIIB
disorientation; extremely LOST all of the time IIIB
cognition; word disorganization II
memory; word loss, poor memory II
skin; terrible break out of reactive cystic acne, face and back IIIB
skin; scalp, dandruff O
anxiety IIIB
depression so intense they sent me inpatient in April 14, MY BIRTHDAY. IIIB
memory problems IIIB (forget how to get places that I've been before, forget where I've put things, CONSTANTLY)
memory; poor word recall II
memory; poor retention of the written word IIIB
eyes; struggle with reading (eyes move around endlessly) [nystagmus] IIIB
cognition; disorganized thought IIIB
speech; difficulty holding conversations IIIB due to memory & poverty of speech
crying response, an overactive IIIB
skin; acne (which is slowly subsiding & gradually lessening) IIIB
fatigue, pervasive, makes keeping a steady schedule pretty impossible IIIB
dizziness II
disturbed sleep IIIB
completely disoriented inside of theater, IIIB despite working there 8 hr. shifts daily for over a week.
confusion; even simple tasks have me fumbling & confused. IIIB
neurological problems IIIB
psychologically symptoms IIIB are ten times worse than initial problem.
frustration, terrible IIIB

depression, severe IIIB
intolerance to starches IIIB
phobias to food [anorexia] I had fasted for 2 straight weeks IIIB
phobia to water [hydrophobia] The only thing I can drink is water, and I have to drink as excruciatingly little as possible in the mornings. The less I can drink before 4:00 PM, the better. Eating or drinking anything else will set off the severe depression.
amenorrhea; I haven't had a period since January, but the depression comes once a month regardless of what I eat then. IIIB
depression; complete despair IIIB praying I don't die.
zaps, brain IIIB
dizziness, light-headedness IA
anxiety IIIB
insomnia IIIB
dreams, intense insane IIIB
panic IIIB attacks every day
paranoia; became IIIB increasingly paranoid when going to sleep constantly feeling like someone was watching me Paranoia stopped after moving back with parents
pain, abdominal II
pain, all over my body IIIB
headaches O
muscle twitching II
tightness around head II
constipation II
diarrhoea, yellow II
anxiety / nervousness IIIB
depersonalisation IIIB
mood swings; emotional rollercoaster
crying, uncontrollable IIIB
shivering, spells of IIIB
nausea II
dizziness IA
shaky; feeling very jittery and shaky inside like I had way too much caffeine [caffeine affects the ear] and this persisted the ntire time I took paxil. IIIB
zaps, brain IIIB
skin; hot spots on my body IIIB
tingling, strange, at the base of my spine which was very uncomfortable IIIB
anxiety attacks (never had them before) IIIB
crying spells IIIB
headaches, topical O
dizziness IA
nausea II
sensitivity to sound, extreme IB
muscle; weak and aching quadriceps II
tremors in my fingers and hands IIIB
ching in my left arm and left side IIIB
fear that what I was experiencing was from a terrible disease; felt that I was dying IIIB
esophagus, spasms in my II
nausea IA
smell, diminished sense of II
dizziness IA
weak and aching legs IIIB
tremors in my fingers. IIIB
neck, strange tight sensation in the back of my II
tongue, a burning sensation on my II
violent thoughts IIIB
brain zaps IIIB
sleep; abnormal, parasomnia IIIB
insomnia IIIB
tremors, night IIIB
intolerance to sudden loud noises (this still continues after 3 years of being Seroxat/Paxil free) IB
agitation IIIB
lack of Empathy IIIB

vision experiences, peculiar IIIB
memory loss, short term IIIB (this still continues after 3 years of being Seroxat/Paxil free)
sleep; irregular sleep patterns, when I wake I can never get back to sleep, no matter how tired I feel.
paranoid IIIB
irrational thoughts IIIB
stress, situational IIIB
depression II
panic IIIB
severe depression IIIB unresolved although some spells of better mood occasionally, not 24/7 hell
anxiety, severe physiological IIIB
panic IIIB
fear; random feelings of horror or terror for no reason (not a panic attack) IIIB
dysphoria, LSD style IIIB
emotional dysregulation, extreme over the top emotions IIIB
anxiety, severe psychological / IIIB
intrusive thoughts / images, rumination IIIB
fears, irrational IIIB
akathesia, moderate IIIB
self-harm, urges to IIIA
panic attacks IIIB
depersonalisation IIIB
derealisation IIIB
suicide ideation IIIA
muscle twitching, constant II
zaps, electric, constant O
obsessive, repetitive thoughts IIIB
temperature, low body IIIB
heart palpitations IIIB
flu-like symptoms, hot / cold, IIIB
shaking / shivering IIIB
eyes; waking up with strobe light effect in my eyes IIIB
images; slideshow of random images in my head IIIB
psychosis; weird dysphoric dream state while wide-awake IIIB (not imagination)
nausea IA
intolerance of stimulants (months 2-5) IIIB
hallucination, audio; music playing in my head (not just having a song stuck in my head) IIIB
intolerance to any kind of stress whatsoever IIIB
memory loss, short term II
confusion, IIIB
cognitive ability, very little IIIB
fatigue, insane spells of IIIB
faintness; periods of feeling faint IIIB
self-concept, obliteration of self-identity IIIB
self-esteem, loss of IIIB
exhaustion II
irritable IIdepressed II
sensitive to noise IB
concentration, poor II
memory, poor II
dreams, vivid IIIB
depersonalization IIIB
dizzy spells IA
vision blurry IIIB
intolerance of alcohol IIIB
eyes; itchy irritation around the eyes not solved by medication given for both allergy and infection II
panic, really severe IIIB
anxiety, depression IIIB
zaps, head, at a very severe level. O
anhedonia IIIB
brain Fog II

motivation, lack of IIIB
sex; no libido IIIB
self-concept, obliteration of IIIB
self-esteem, loss of IIIB
anxiety, severe (for no apparent reason) IIIB
appetite, no IIIB
cognitive abilities, impaired IIIB
crying jags (for no apparent reason) IIIB
depression, major IIIB (for no apparent reason) *NOTE:* I have **NO** history of depression or anxiety
insomnia IIIB
memory, impaired II
suicide ideation IIIA
abdominal pains II
acid reflux II
agoraphobia IIIB
akathasia IIIB
igraines O
diarrhoea chronic and severe. This continued daily for 2 ½ years
stomach pain, terrible, and heartburn that burns all the way up to my throat. I'm afraid I'll end up with bleeding ulcers
homicidal thoughts IIIA
suicidal thoughts IIIA
aggressive thoughts IIIB
unusual thoughts
mood changes
mental changes IIIB
mania, induced IIIB
agitation IIIB
restlessness
insomnia IIIB
sex; unable to sustain an erection or ejaculate II
gut; black stools II
fatigue, unusual II
headache at the back of the head II
mouth, dry mouth IIIB
vision, blurred vision IIIB
numbness
weakness IIIB
heartbeat, slow IIIB
energy, feeling of no IIIB
crying fits IIIB
akasthasia IIIB
dreams, agitated, about knives and guns IIIB
nightmares IIIB
derealisation IIIB
depersonalisation IIIB
agoraphobia; no desire to leave the house IIIB
fear; random horror and terror (was not a panic attack) IIIB
tingling, prickly sensations II
nerves; pins and needles II
painful sensations II
headaches, fuzzy sharp O
vision; floaters, visual disturbances (all checked by optician and normal) blue flashing lights IIIB
pain in fingers, feet, resembling nerve pain IIIB
self-identity, loss of IIIB
strange thoughts still persisting IIIB
temperature sensations (hot and cold) IIIB
mouth, burning sensations on II
brain, feel as though part of my brain is missing IIIB
sensitivity to sound/IB
sensitivity to light IIIB

confusion/ IIIB
disorientation IIIB
disequilibrium, strange balance IA
memory problems II
breathing; feelings of lack of oxygen IIIB
tingling, numbness IIIB
painful sensations IIIB
nerves; pins and needles (chest, fingers, hands, arms, you name it) IIIB
panic attacks IIIB
anxiety IIIB
extreme OCD IIIB
insomnia, severe IIIB
back pains and stiffness II
lack of appetite, dried mouth II
appetite, increased
salivation, extreme II
stomach pains II
crying spells IIIB
weight loss IIIB
depression IIIB
headaches and head pressures O
obsessive thinking about dying and fear of dying IIIB
intolerance to heat IIIB
muscle pains II)
anger, extreme (since 2nd month until 4th month post Paxil) IIIB
anger IIIB
sensitivity to loud sound IB
intolerance to stress IIIB
concentrating II
memory, significant short-term / working memory issues II
speech; problems with verbal fluency II
memory; problems with word recall II
cognition; difficulty reading II
concentrating, difficulty focusing II
zaps, accompanied by dizziness O
dizziness IA
skin sensitivity, extreme IIIB
rashes IIIB
skin; cystic acne eruptions
skin; dry itching scalp
crying spells IIIB
eyes; floaters/ IIIB
visual disturbances IIIB
tinnitus IB
sleep; hypersomnia IIIB
tremors in right hand II
muscle twitches II
flushing; hot flashes IIIB
tinnitus; ringing in my ears IB
suicide, two IIIA, IIIA attempts in first few weeks on the drug,
lethargic, very IIIB
irritability II
digestive II
stress, susceptible to IIIB
irritable, angry spells IIIB
zaps, brain O
chest pain II
nausea and vomiting IA
teeth, lost two more teeth two weeks ago (that makes 8 that have fallen out since being on Paxil O
intolerance to many foods IIIB

sleep disturbance IIIB
eyes; cannot watch a movie II
emotion; inability to laugh IIIB
disconnect; lack of awareness as to how moods affect others IIIB
homicide III
zaps O
whooses IIIB
fatigue IIIB
severe pressure / pain in the head O
neck, pain IIIB
pains, sharp, in one part of my head that come and go IIIB
blood pressure; supine (postural) blood pressure changes (some severe) IIIB
anxiety, off the charts IIIB
flushing; hot flashes to where I have to step outside in the winter IIIB
sleep; inability to go back to sleep (feeling TOO awake) IIIB
heart palpitations (PVCs/PACs) – actually charted on EKG IIIB
atrial fibrillation, one episode of AFIB IIIB
dizziness IA
visual blurriness IIIB
vision, increase in flashers / floaters IIIB
chest, severe tightness of IIIB
breathing; shortness of breath that comes and goes IIIB
tinnitus, ringing in the ears IB
blood pressure; major orthostatic intolerance (i.e., significant pulse AND blood pressure decrease / increase upon lying or standing similar to POTS IIIB
mood swings IIIB
anger not reasonable to the situation IIIB
intolerance to stress; ZERO tolerance to anything stressful IIIB
zoning out IIIB
speech; not as articulate as I once was IIIB
confusion; cannot communicate thoughts as clearly IIIB
numbness and/or tingling in all parts of the body either separately or all together IIIB
stomach cramps, intense, lasting for 2–5 hours IIIB
visual jumps or what I called shifting IIIB
crying jags IIIB
insomnia IIIB
dreams, vivid IIIB
weight loss, dropped 44 pounds in 4 months, which left me at 99 lbs IIIB
vomiting after a meal, constant daily IIIB
hair loss; hair fell out at a rapid rate and is not [now] permenantly thinner IIIB
depression IIIB
rages, sudden IIIB
muscle aches II
movement disorder IIIB
sensory disturbance IIIB
loss of stamina IIIB
tachycardia IIIB
personality change IIIB,
loss of concentration, II
loss of memory IIIB
insomnia IIIB
anxiety attacks with vitamin use. IIIB
less muscle energy II
mentally foggy and lazy IIIB
heaviness IIIB
sleep, non-restful sleep IIIB
socially withdrawn IIIB
distressing changes in physical appearance IIIB
derealisation IIIB
depersonalisation IIIB

irritability IIIB
vision problems, foggy vision, focus problems IIIB
cognitive problems, IIIB
foggy brain, IIIB
memory and IIIB
concentration problems II
sexual obsessions IIIB
muscle weakness, II
fatigue, IIIB
low stamina IIIB
apathy, IIIB
emotions, lack of [flat affect] IIIB
balance problems IA
muscle ache II
skin, sticky skin, waxy hair, IIIB
skin rashes IIIB
head, horrible sensations like heavy head O
head pressures O
intolerance to stress, can't handle stress IIIB
anxiety, worse anxiety than pre the AD's--its a physical anxiety IIIB
dizziness in different ways, vertical and horizontal IA
sensitive for [to] sound IB
anhedonia, inability to enjoy everyday things, IIIB
inability to enjoy music, IB
inability to watch movies
personality change IIIB
cognition; I just can't think abstractly anymore IIIB
cognition, I can't learn. IIIB
cognition, my attention to detail sucks. IIIB
head, burning/frying feeling in head IIIB
cognition; foggy / cloudy thinking IIIB
sensitive to lght IIIB
tinnitus; ringing ears IB
flat affect; blunt emotions IIIB
memory loss, short term II
concentrate, inability to IIIB
dizziness / vertigo IA
nausea IA
diarrhea IIIB
constipation IIIB
tinnitus IB
hyperacusis [sensitivity to sound] IB
weight loss IIIB
appetite, loss of IIIB
depression IIIB
anxiety, extreme IIIB
concentrate, unable to II
restlessness, inner IIIB
panic attacks IIIB
muscle twitching / jerks II
insomnia IIIB
waking up disoriented in the middle of the night IIIB
crying spells IIIB
irritability / feeling angry IIIB
zaps, head O
dizziness; lightheadedness IA
flu-like symptoms IIIB
skin; acne IIIB
pain, heel pain and swelling O
tinnitus IB

zaps, head, mild and occasional O
muscle twitching / jerking, mild II
hair loss; falling hair IIIB
skin; acne IIIB
bladder problems IIIB
diarrhea IIIB
constipation IIIB
nausea IA
sleep, disrupted IIIB
obsessive thoughts IIIB
anger, extreme IIIB
memory, short-term, loss of; forgetfullness II
intolerance of stress; get stressed out easily IIIB
emotionally numb IIIB
emotional, extremely IIIB
anhedonia; do not enjoy nothing IIIB
sex, extreme sex drive IIIB
symptoms, too many other to list here. IIIB x 10
emotional numbness IIIB
emotional, extremely IIIB
anxiety (never had before) IIIB
suicidal thoughts IIIA
headache, chronic IIIB
tics IIIB
breathlessness IIIB
skin, very dry skin IIIB
stress, inability to cope with stress IIIB
stiffness IIIB
restless legs IIIB
fatigue; very tired IIIB
temperature, low body IIIB
teeth, grinding on teeth IIIB
rage IIIB
mood swings IIIB
flu-like symptoms IIIB
lethargy IIIB
suicidal, I took overdose with codeine. IIIA
parkinsonianism, unable to coordinate in right direction IIIB
co-ordination, extreme lack of, IIIB
mania IIIB
agitation IIIB
suicidal intent. IIIA [attempt]
cognition, loss of 30 points off I Q IIIB
speech; IIIB
cognition; muddled thinking IIIB
speech; communication difficulties, more difficulties verbally than written IIIB
cognition; delay in processing. IIIB
dizziness IA
blurred vision IIIB
balance, loss of IA
agitation IIIB
frustration IIIB
OCD behaviours did not have before IIIB
breathlessness on exertion IIIB
fatigue, chronic IIIB
tremors IIIB
tics IIIB
epilepsy, episodes of petit mal IIIB
numb legs and feet/ IIIB intermittent, sometimes so bad I can't move them.
anxiety, social IIIB

memory, disfunction IIIB
inability to organize IIIB
muscle; seized up
tremors, severe IIIB
autism spectrum IIIB
sensitivities to chemicals, supplements, medications, herbs, caffeine, etc. IIIB
depersonalization, IIIB
emotional numbness [flat affect] IIIB
nerves; reactions, stunted IIIB
nerves; slowed reaction times IIIB
spacing out, zoning out IIIB
absentmindedness II
brain fogginess, etc.
cognition; trouble with elaborating complete sentences IIIB
speech problems IIIB
speech, tone in, a monotone blahh, single word answer type of speech IIIB
anxiety, social IIIB when in the presence of others I freeze up, feel butterflies, mouth and throat close, IIIB
feel disquietingly quiet, shy, timid, sensitive, loner,
desire to be left alone IIIB used to enjoy others company.
periods of impending doom, IIIB
anxiety unrelated to events, obsessive anxiety IIIB
balance, unco-ordinated walk, feel light IIIB on my feet, like the wind can tip me over.
fatigue, easily fatigued without even exerting myself. IIIB
strength, decreased physical activity II
memory problems, IIIB
cognition, lack of common sense, IIIB
poor driving IIIB
insomnia, periods of IIIB
racing thoughts; too many thoughts to manage IIIB
suicidal depression, like i want to end it all. IIIA
cognition, loss of familiar learning IIIB
suicidal thoughts IIIA
feelings of severe burning in hands and IIIB
feelings of burning sweaty feet with heavy tingling IIIB
head, bad pressure in head IIIB
head, crawling pains in IIIB
muscle tics II
lights flashing in eyes IIIB
skin, sensitive IIIB
burning sensations on arms, legs and face IIIB
hallucination, audio; Someone talking in my head, IIIB but it's not me.
fear of EVERYTHING, IIIB
absolutely convinced I am going to die IIIB EVERY SINGLE DAY.
obsessive thinkingIIIB
obsessive talking IIIB
anxiety, much worse then I ever had before I took this drug. IIIB
obsessive thoughts mixed with anxiety, worse in the morning, I wake up to IIIB
memory, very poor IIIB
concentration, very poor IIIB
cognition; thoughts are hard to put together, IIIB
dyslexia; I have noticeable trouble with spelling words, not present before IIIB
sensitive to other medications IIIB
spacey like I'm stoned or something IIIB
vision, spacey IIIB
head, dull ache in O
feet still burn and tingle some IIIB
spaceiness IIIB
depression, on and off IIIB
thinking, problems IIIB
concentration, improvingIIIB

parasthesias (burning, tingling) IIIB
myoclonus IIIB [muscle spasms]
facial twitches II
akathisia (despite 3 medications to control it) IIIB
intolerance to stress IIIB
cognitive deficits, severe IIIB
cognition; poor reasoning skills, IIIB
inability to organize; inability to multitask, IIIB
imagination or creativity, almost no IIIB
memory loss, severe IIIB
cognition; learning problem; inability to retain new information IIIB,
comprehension skills, poor IIIB
confusion IIIB
memory problems; skewed recall of facts IIIB
cognition: sequential thinking, etc. IIIB
digestion, poor; occasional passing of food completely undigested (in spite of enzyme supplementation) IIIB
hypochlorhydria IIIB
feelings; often do not feel satiated after large meals IIIB
intolerance to certain foods IIIB
digestive enzyme deficiency; almost all of the foods I can tolerate must be cooked before eaten
anhedonia IIIB
rage, bouts of extreme IIIB
suicidal thoughts, daily IIIA
depression IIIB
emotional blunting, long bouts of extreme IIIB
paranoia, bouts of IIIB
skin; acne IIIB
skin, dry IIIB
inability to gain weight IIIB (am 105-110 lbs. and 5'4)
menses; lengthened period IIIB
leg aches not associated with exercise IIIB
sleep, poor IIIB
panic, feelings of, which can last for weeks IIIB
temperature, inability to gauge IIIB
time, inability to estimate what time it is IIIB
dyslexia IIIB, II
intolerance to cold IIIB
headaches while driving IIIB
fatigue, extreme; exhaustion from normal activities IIIB of daily living
muscles, atrophied II
thyroid, slightly overactive IIIB
sensitivity, extreme, to artificial light IIIB
sensitivity, extreme, to sound IB
adrenaline rushes IIIB
panic attacks IIIB
leg aches IIIB
parasthesias IIIB
memory problems (amnesia) IIIB
cognitive problems IIIB
organizational skills, poor IIIB
comprehension skills, poor IIIB
concentration, poor IIIB
memory; difficulty retrieving words IIIB
cognition, reasoning skills, poor IIIB
cognition, almost no critical thinking skills IIIB
imagination / creativity, poor IIIB
dissociation IIIB
akathisia, intermittent "mild" IIIB
tinnitus, intermittent IB
taste and smell, poor sense of IIIB

time, poor sense of IIIB (feel like I live in the moment)
temperature, poor sense of IIIB
sweat, barely
apathy, severe IIIB
anhedonia IIIB
sleep, frequent wakenings IIIB poor quality sleep
cognition; general decrease in awareness IIIB
heart occasionally feels like it is speeding up,skipping beats IIIB
fear (lessening over time) IIIB
ears; ears are straining to hear, not related to any outside stimulus IB (this starts abruptly and not because I am trying to hear omething)
unusual thirst; very thirsty some days, unrelated to outside temp IIIB
bladder; foul smelling urine IIIB
jittery feeling IIIB
muscle weakness, feelings like I can't easily make a fist, II
muscle, weakness, feeling like I can't type
speech; feeling like I can't talk (intermittent) II
vision problems--blurry eyes, IIIB
eyes; poor dark adaptation IIIB
vision; perceptual difficulties, seeing what I am told are "tracers" IIIB (intermittent)
integration, cognitive; difficulty scanning a large number of objects to find the one I want IIIB
integration, visual IIIB my brain does not appear to be creating a "whole" out of the parts of the picture
bowel problems, intermittent IIIB
dizziness (occasional) IA
pain on left side of bowel IIIB (occasional)
anger, extreme, during my period IIIB
skin, dry IIIB
skin, acne IIIB
hair loss IIIB
bloating
weight gain IIIB
feel "flat" on cloudy days IIIB
irrational IIIB
manic-IIIB
depressive person IIIB
psychotic IIIB
depressed, not too happy IIIB
anhedonia - I don't enjoy ANYTHING I used to enjoy. I merely exist from one day to the next IIIB
emotional pain; every second of every day now is absolute Torture IIIB
emotional stuff IIIB
physical pain IIIB I
suicide, thoughts of, and death IIIA
depression
anxiety, high
body pain.
angry and bitter, IIIB
drug addict IIIB
no direction or hope in life.
low estrogen, no periods which have returned to normal after many months II
memory problems IIIB
disorientation IIIB
vertigo, severe IA
agoraphoby IIIB
stiffness, walk problems IIIB
muscle weakness II
anxiety, major IIIB
depression, severe, IIIB
suicidal ideas IIIA
motivation, absolutely no, for everything II
dizziness IA

anhedonia II
appetite, loss of IIIB
diarrhoea II
back pain IIIB
fatigue, big IIIB
sleep, poor quality IIIB
anger, irrational IIIB
aggressiveness IIIB
memory problems IIIB
sex; libido loss completely IIIB
delayed / no orgasm IIIB
sex; the orgasm is horrible IIIB
emotional numbness IIIB
depression IIIB
insomnia IIIB
restless leg syndrome (sometimes very painful) IIIB
back pain IIIB
anxiety, uncontrollable IIIB
emotional outbursts IIIB
sensitivity to supplements, medication, processed foods, additives chemicals, etc. IIIB
anxiety IIIB
alienation from my emotions IIIB
emotions; flattened affect IIIB
cognitive, slowed thinking IIIB
sleep, poor, bouts of , IIIB
premature ejaculation, alternating IIIB
fatigue after consuming sweets IIIB
moods, profoundly low
muscle spasms, neck II
nerves; a stitich-like contraction feeling that runs down my face II
nerves; a stitich-like contraction feeling that runs down right side of body. II
akasthasia, terrible, IIIB
suicidal urges IIIA
cognitive and physical symptoms IIIB x 10
zaps, electric shocks in my head IIIB
insomnia. IIIB
fatigue; very tired all the time, every singleday. IIIB
mood swings, sometimes 'happy' other times crying, often within a few minutes. IIIB
memory problems, short-term IIIB
confusion and 'brain fog' IIIB
sensitive to bright lights IIIB
sensitive to sound, especially sudden sounds IB
which can send like electric shock through my whole body IIIB
sexual dysfunction; low sex drive and impotency IIIB
temperature; feeling cold inside yet hot and sweating on the outside as though I have the flu – but as far as I know IIIB one cannot have the flu for eight years!
muscle spasms, very painful, like pulsating cramp II
skin, itching under the skin, anywhere on my body. IIIB
numbness on the surface of the skin. IIIB
nerves, pins and needles anywhere but most often in my hands and arms. IIIB
twitching; hands or feet sometimes twitch or move by themselves. IIIB
vision; sometimes *I cannot see properly!* IIIB
cognition; lowered IQ IIIB
co-ordination, cannot write by hand
cognition; cannot do maths IIIB
anxiety, severe IIIB
depersonalization IIIB
mood swings, severe IIIB
brain fog IIIB
vision, blurred IIIB

exercise intolerance IIIB
tremors IIIB
nausea IA
head pressure O
orgasm intolerance (causes brain fog) IIIB
agitation IIIB
severe nerve pain IIIB
muscle pain II
dizziness IA
memory problems IIIB
concentration problems II
eyes; floaters in IIIB
exercise intolerance IIIB
headaces O
nausea IA
anxiety IIIB
mood swings IIIB
brain fog IIIB
stiffness IIIB
muscle weakness II
sweating IIIB
akathasia, extreme IIIB
breathing; couldn't breath properly had to think about every breath otherwise i might stop. IIIB
insomnia IIIB
sleep; weird head rush every time i try drift to sleep, along with shot of adrenalin to my heart IIIB
awakening up to adrenalin rushes IIIB
hallucinations, audio; music playing in my head IIIB.
awakening; s[l]ideshow of scary images when i woke from my sleep IIIB (extremely scary).
hallucinations, visual; keep thinking i see things that arn't there IIIB
heart pounding 24/7. IIIB
vision, blurred IIIB
dreams, vivid, that felt real IIIB
head felt vacant. IIIB
emotionally dead. IIIB
depression, chemical, (everything and everyone looked dead) IIIB
chest pains and tightness IIIB.
abdominal pains, extreme II
depersonalization/ IIIB
derealisation. IIIB
dark horrible images when i close my eyes. IIIB
fear, intense, for no reason. IIIB
hiccups for 2 days (would start as soon as i got rid of them) [vagus nerve?] II
teeth grinding IIIB
paranoid IIIB
tinnitus IB
whooshing sound in my head IB
hallucination, audio; cricket chirping noise IB
anorexic IIIB
racing thoughts IIIB
concentrate, can't at all II
memory, can't remember anything IIIB
heart constantly pounding IIIB
heart palpitations IIIB
akathasia IIIB
restless leg syndrome. IIIB
anhedonia, chemical can't feel any love / joy. IIIB
muscle twitches everywhere. IIIB
cognition, feels like something is missing from my brain IIIB
chest tightness IIIB
anxiety, mild II

vision, blurred IIIB
memory, short-term IIIB
cognition; use the wrong words constantly. II
head rushes. IIIB
heart palpitations IIIB.
concentration, lack of II
head pains / sensations O
head pressure, burning / sizzling, O
headaches
head, stabbing pain in
cognitive, every problem there is IIIB, IIIB, IIIB
tinnitus IB
involuntary movements IIIB
chills (havn't had for months IIIB)
depersonalisation/ IIIB
derealisaation IIIB
dizzy spells IA
sensitive to light IIIB
sensitive to sound IB
flat affect; blunt emotions IIIB
other symptoms IIIB IIIB IIIB
depressed. IIIB
mute. IIIB
spacy. IIIB
dreams, vivid IIIB
sleep disturbance. IIIB
memory problems. IIIB
sleep; unable to get good delta, deep sleep and feel recharged.
emotionally numb.
emotion, really sensitive to IIIB
sensitive, especially to music. IB
memory, zero IIIB
brain damage IIIB
cognitve problems, more IIIB, IIIB
arthritis, UCTD disease. which causes arthritis
fatigue, chronic IIIB X 10
immune system disorder IIIB
bowel; IBS IIIB
agitation IIIB
akathisia, IIIB
skin; head-toe itching. IIIB
weight gain IIIB
sexual disfunction IIIB (inability to orgasm)
apathy IIIB
flat affect; no normal emotions IIIB
nausea IA
nightmares IIIB
diarrhoea IIIB
emotional instability (tearfulness) IIIB
anxiety, chronic IIIB
restlessness IIIB
numbness in mouth / teeth IIIB
headache, IIIB
pressure on brain O
muscle spasms II
weight loss (I've lost 1.5 stones since beginning the withdrawal process) IIIB
digestive problems, extreme IIIB
diarrhea, ongoing bouts of IIIB
dizziness, generalized IA
nausea IA

headache, migraine O
fatigue, extreme II
fatigue; poor stamina during physical activity IIIB
muscle and joint pain IIIB
hip joints pain IIIB
muscles spasms II
back pain, specific, thoracic spine II
back pain, lumbar muscles II
vertigo, episodes of IA
sleeping, difficulty IIIB
concentrating, difficulty recalling even simple words II
nerves; "pin prick" feelings in my right cheek II
nerves; "pin prick" feelings in my right cheek and right forearm simultaneously IIIB
pressure feeling in my face II
grinding teeth during the night IIIB
waking up sore from head to toe IIIB
muscle and joint pain, minor IIIB
digestive issues II
fatigue, intermittent II
Parkinsonism...the worst of any symptom I have had. IIIB
speech; inabilty to articulate, II
jaw clenching, II
grimacing... II
mouth movements, involuntary II
co-ordination; chewing is even difficult as my mouth and tongue do not seem to be in sync. II
saliva, excessive II even with dry mouth
mouth, dry II.
tinnitus, daily screechingIB
weigh gain, 22 pounds IIIB.
cognitive; indifference to smoking and drinking IIIB, IIIB
euphoria; feeling freed from the dark cloud IIIB that held me down
anxiety IIIB
dizzy spells IA
flu like feelings IIIB
concentration, poor II
memory, poor short-term II
anxiety IIIB
joint aches and pains IIIB
sensitivities to many and mainly acidic foods IIIB
nightmares, horrific IIIB
panic attacks IIIB
anxiety IIIB
restless legs IIIB
memory problems IIIB
insomnia IIIB
depression IIIB
jaw clenching II
joint and muscle pain II
zaps, brain IIIB
vision, improved II
twitching II
gut; IBS IIIB
zaps, brain waves IIIB
sound; cracking up the neck and in the head, IIIB
tinnitus, IB
insomnia, IIIB
vision, blurred IIIB
flat affect; emotional bluntness IIIB,
apathy; no interests in anything II
anxiety, raging IIIB

intolerance to music IB
intolerance to watching movies IIIB
intolerance to reading anything during withdrawal IIIB
mood swings, IIIB
zaps IIIB
fatigue IIIB
anxiety, moderate II
stomach issues II
appetite, loss of II
anxiety, extreme IIIB
symptoms, other common withdrawal II
agoraphobia IIIB
symptoms, other withdrawal IIIB
heart issues IIIB
anxiety
cognition; reading: I used to be a quick reader and now I struggle with comprehension II